APUN
The Arctic Snow

Teacher's Guide

APUN
The Arctic Snow

Teacher's Guide

Matthew Sturm

University of Alaska Press
Fairbanks, Alaska

University of Alaska Press
P.O. Box 756240
Fairbanks, AK 99775-6240

Library of Congress Cataloging-in-Publication Data

Sturm, Matthew.
Apun, the arctic snow : teacher's guide / by Matthew Sturm.
p. cm.
ISBN 978-1-60223-070-5 (pbk. : alk. paper)
1. Snow—Arctic regions—Study and teaching (Elementary) I. Title.
QC926.32.S78 2009
551.57'842113—dc22

2009016360

Text and cover design by Paula Elmes, ImageCraft Publications & Design

Cover and chapter divider illustrations by Ken Libbrecht

This publication was printed on acid-free paper that meets the minimum requirements for ANSI / NISO Z39.48-1992 (R2002) (Permanence of Paper for Printed Library Materials).

Dedicated to

Arnold Brower, Sr.,

of Barrow, Alaska,

who understood snow.

Arnold Brower, Sr., and the author in 2008. (Photo by Craig George.)

Contents

About This Book

I wanted to write a simple book about arctic snow. It is a fascinating snow cover, absolutely essential for human, plant, and animal life in the Arctic. Yet it is so common and is present for so long each year that it is taken for granted by many residents of the Arctic. This is particularly true of modern arctic kids, whose survival no longer hinges on understanding the snow. My hope is that this simple book will help rekindle interest in arctic snow and remind both kids and adults about the special and important place the snow has in their lives. For non-arctic residents, I wanted to describe a type of snow it has been my privilege as a scientist to study for the past twenty-five years, a type that is probably quite different from the snow they are familiar with but one that is definitely worth getting to know.

I also wanted to write a book that introduced kids to Iñupiaq words for snow. The Iñupiaq culture is so enmeshed in the world of snow and ice, it should come as no surprise that the language is rich in terms describing these features. Lists of snow terms have been available for many years, and there has been some controversy about just how many words there are. Despite this attention, when I prepared this book it became clear that there was a "fuzziness" in many of the existing definitions. The reason for this may be that linguists, not snow experts, compiled the original lists, and some of the nuances of Iñupiat snow knowledge may have been lost. To refine the list, I have been conducting interviews with Native elders in Barrow for the past five years, a process that is still underway. I am particularly grateful to Arnold Brower, Sr., Kenny Toovak, Fannie Akpik, and Martha Stackhouse for sharing their wisdom about snow and about Iñupiaq with me. The words that appear in the book are based largely on my interviews with them. I also thank Glenn Sheehan of the Barrow Arctic Science Consortium for facilitating these interviews, and Dr. Larry Kaplan, director of the Alaska Native Language Center at the University of Alaska Fairbanks, for checking the spelling of the words.

This little book is a work in progress. As a scientist, I continue to learn more about the snow each year. As an amateur linguist, each interview I conduct expands my appreciation of the richness of knowledge about snow that exists in the Iñupiat culture.

—Matthew Sturm

Acknowledgments

Many people have helped make this book possible. Larry Kaplan, Arnold Brower, Sr., Kenny Toovak, Craig George, Martha Stackhouse, and Fanny Akpik helped with the Iñupiaq glossary. Glenn Sheehan of the Barrow Arctic Science Consortium supported the work in many ways. Carl Benson, Sam Colbeck, Chuck Racine, and Don Perovich taught me much about snow. Jon Holmgren, Walt Tape, and Ken Libbrecht were particularly generous with their snow crystal photographs. Sue Mitchell and her production team at the University of Alaska Press made the book look beautiful. Finally, my wife, Betsy, tested the book on her second-grade class and has always encouraged my writing and drawing.

Snow and Life

Barrow, Alaska

Barrow is a good place to learn about the arctic snow cover. In this sketch map of Barrow, Alaska (the town), and Point Barrow (the northernmost point in the United States), the cloud on the right is over the Beaufort Sea, the one on the left over the Chukchi Sea. The lagoon south of Point Barrow is called Elson Lagoon. The jet landing at the airport has approached the runway from the sea. Passengers would have had a good view of the sea ice on their left as they landed.

Barrow, Alaska, is in the Arctic. What is the Arctic? It is the part of the Earth north of the Arctic Circle, which is at 66° 33' north latitude. The circle is defined by the seasonal movement of the sun. Above the Arctic Circle, the sun sets in the winter and stays down for months at a time. It rises again in spring and stays up most of the summer. This annual cycle makes for dramatic differences between winter and summer temperatures. It also leads to a large difference between winter, when snow blankets the ground, and summer, when plants grow and it is green. *Apun* is the story of that long-lasting blanket of snow. The connection between the solar cycle and the nine-month snow season is summarized in Figure 1.

Geographically, the Arctic consists of an ocean centered on the North Pole (the Arctic Ocean) surrounded by a ring of land. The pan-Arctic

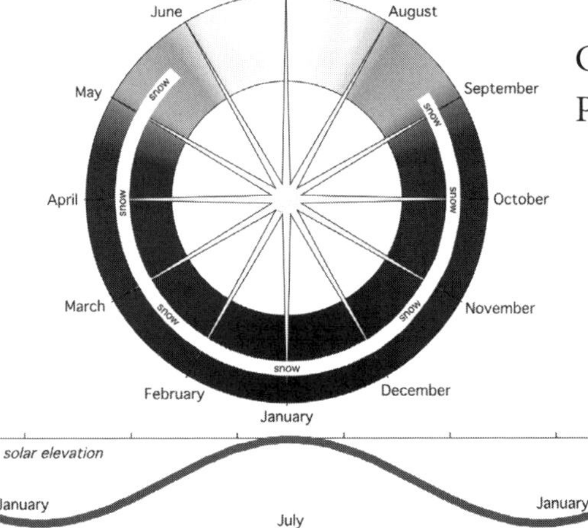

Figure 1: The annual arctic cycle of light and darkness, snow duration, and growing season length. The shading of the wheel suggests the available daylight, with twenty-four hours of continuous daylight in July and twenty-four hours of continuous darkness in January. The white line in the circle indicates the months when snow is present on the ground. The gray line at the bottom suggests the annual waxing and waning of solar intensity, peaking in July and dropping to zero in late December and early January.

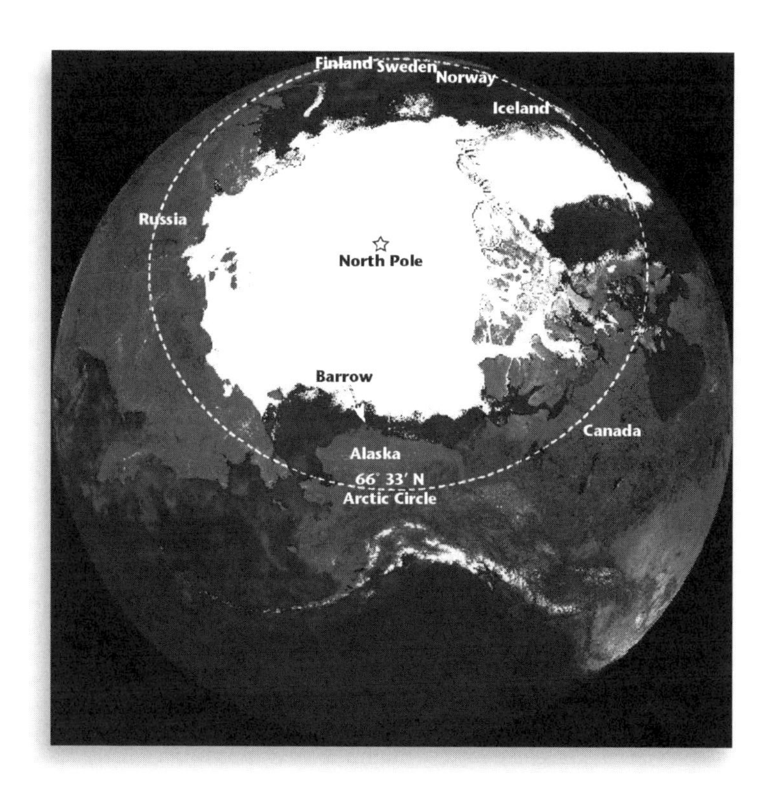

Figure 2: A view of the Arctic from space. The North Pole is in the middle of the Arctic Ocean, here covered by sea ice (white). Barrow is on the Arctic Coast. The snow-cover story told here applies in Barrow and to most other places north of the Arctic Circle (dashed line), including on the sea ice. (Photo courtesy of NASA.)

countries that make up the land are: the United States, Canada, Greenland (administered by Denmark), Iceland, Norway, Sweden, Finland, and Russia (Figure 2). The arctic landmass is covered by vast spruce forests called taiga, rolling tundra plains, and thousands of lakes. Unlike the Atlantic and Pacific oceans, the Arctic Ocean freezes over in winter: by March each year about 95% (4 million square miles or 12 million square kilometers) is covered by sea ice (salty ice formed by the freezing of seawater). Most of this ice melts in summer. In 2007, the current record minimum, only 1.5 million square miles (4 million square kilometers) of ice were left at summer's end. As the climate continues to warm, this residual amount is expected to decline even more. The newest projections suggest the Arctic Ocean will be ice-free in summer within the next twenty years, threatening marine mammals like polar bear, who need the ice.

There are about four million people living in the Arctic, mostly in small towns and villages. The only U.S. state that includes arctic lands is Alaska. About 15,000 Alaskans (out of a total of 670,000) live north of the Arctic Circle. The farthest north point of land in the U.S. is not actually the town of Barrow, but Point Barrow, a narrow sandy spit of land that

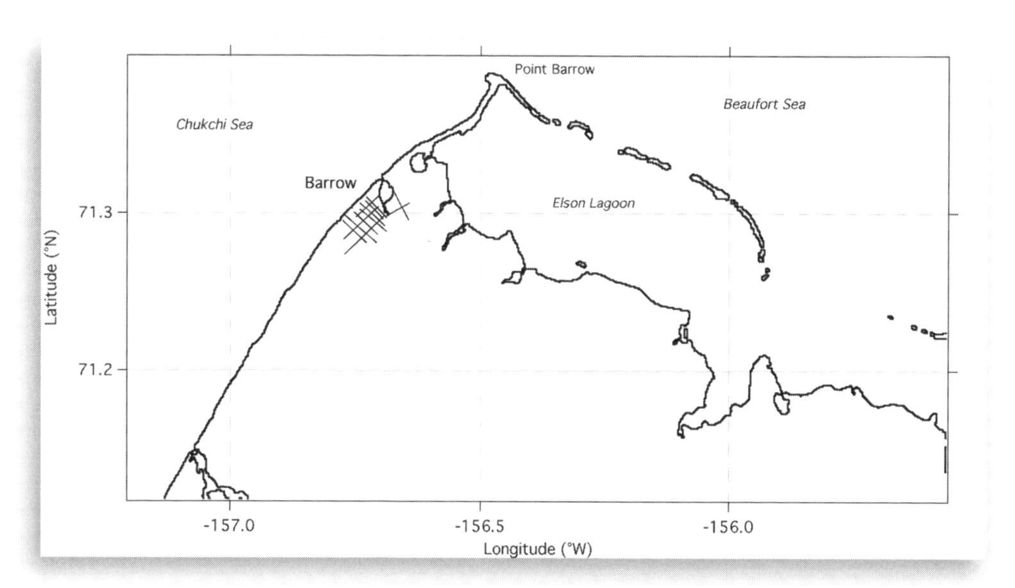

Figure 3: The town of Barrow, Alaska, with Point Barrow, the farthest north point in the U.S., projecting out between the Chukchi and Beaufort seas.

looks out on two frozen seas (the Beaufort and Chukchi seas: see Figure 3). As recently as one hundred years ago, people lived at the point, subsisting by hunting whales and seals, but now nothing is left there but sea ice, driftwood, a gravelly beach, and a tall pole on a cement base. The 4,500 people of Barrow live about 10 miles (16 km) away.

Barrow is a busy place. Barrow residents do all the normal things people do elsewhere in the U.S. They go to work and school, watch TV, play basketball, and hang out with their friends. Barrow has a big supermarket, Mexican and Japanese restaurants, and a modern jet airport. Still, some things are different about Barrow. People there go whaling from the sea ice, hunt caribou and geese out on the tundra, ride snowmobiles for fun and work, and hunt seals. There are other differences from the rest of the U.S., too. For example, most of the year it is hard to tell where the land ends and the frozen sea begins, since both are covered with snow. Blizzards are common, and for sixty-five days in the winter (November 18 to January 22) the sun never comes above the horizon. On the shortest day of the year (December 21) there is just enough twilight at midday to read a book outside, but only for about an hour.

The Snow Calendar

The arctic snow year begins in late September or early October and extends until late May or early June, more than nine months of the year. Despite being long-lasting, the snow cover that builds up is thin, usually less than 20 inches (50 cm) deep. With only five to eight storms per winter, it has relatively few layers. These layers change with time in a process called snow metamorphism. Eventually the winter ends and the weather warms enough that the snow starts to melt. That too is a form of metamorphism, although it ends in the disappearance of the snow cover. The summer is short, though, and soon the snow is back again.

Soon it will begin to snow. Temperatures will drop and the long winter night will start. Through the night, layer by layer, the snow cover will grow deeper. The snow crystals in each layer will change as the winter passes. The sun will come back in January, and by May the snow will start to melt. The brief summer will begin, but the snow won't be gone for long because next October the cycle will start again. This is the life-cycle of the arctic snow cover—*apun* in Iñupiaq.

In northern countries, the importance of the "snow calendar" is reflected in the fact that the year has really only two seasons, summer and winter, separated by two critical transitions known as freeze-up and breakup, which are terms that come from snow and ice. Freeze-up comes in September. At freeze-up, lakes and rivers freeze, and precipitation begins to fall as snow not rain. The snow quickly covers the taiga and tundra, turning them white. Finally spring arrives in May and the snow melts, running off in rivulets, creeks, and rivers. As the water in the creeks and rivers rises, the winter ice cover, which until then has remained intact, cracks into ice floes that are carried down the swollen waterways in a chaotic fashion. This is why the transition is called breakup. What then is the "real" color of the Arctic? White, for snow. Figure 4 shows some real snow calendar data, including the air temperature, the snow depth, and the temperature of the soil. It is data that we collected at a weather station in the Arctic.

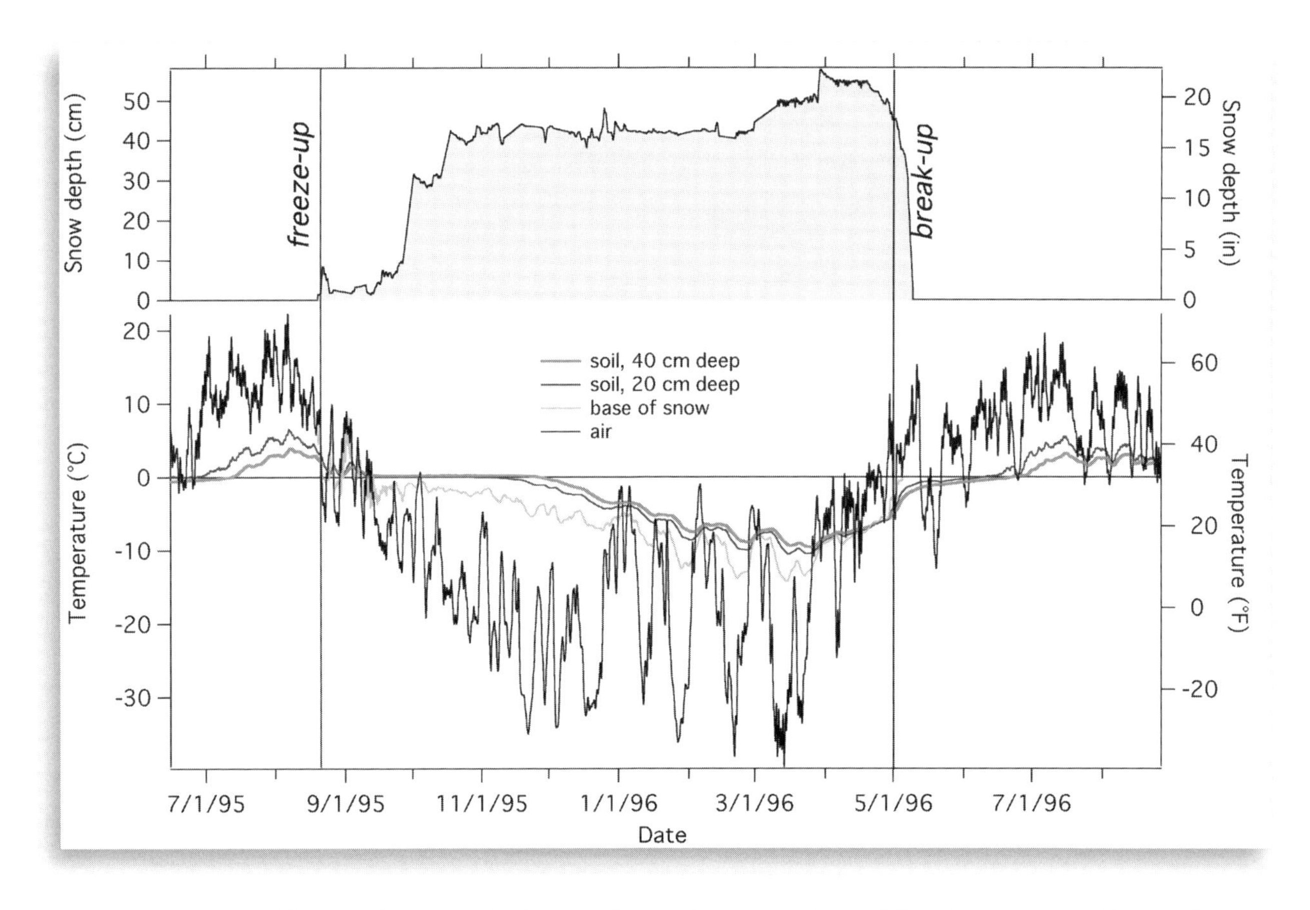

Figure 4: Much can be learned from this record of snowfall and temperature from a site just north of the Brooks Range in Alaska called Imnavait Creek. Because arctic storms are more common in autumn and early winter than later, most of the snow accumulates early, reaching nearly its peak depth by early December. The air temperature (black jagged line) drops below freezing in early September at the same time that snow begins to accumulate. The snow provides good insulation, keeping the soil from fully freezing until nearly January, by which time the air temperature has reached a frigid −31°F (−35°C). At times the ground is 54°F (30°C) warmer than the air. The air temperature rises above freezing in May and the snow melts rapidly, but because the ground remains frozen during the snow melt, little water actually soaks into the ground. The approximate time of freeze-up and breakup are marked by vertical lines. The period between these two lines, a full nine months of the year, is the time of *apun*, the arctic snow cover.

Snow Is a Quilt

The snow cover is important in many ways. It reflects sunlight, which helps keep the Arctic a cold place. It insulates the ground like a quilted blanket, which keeps the plants, and little creatures that live under the snow, warm because they are not exposed to the full force of frigid winter temperatures. It is also a major source of fresh water, and it is a friend to arctic travelers, allowing heavy loads to slide along easily and quickly on sleds. Here, a lemming (which lives under the snow) is enjoying how the blanket of snow, the *apun*, helps keep her warm during the winter.

Apun can mean life or death for the plants and animals of the Arctic. For example, the snow cover is the blanket that keeps the lemmings warm in the winter, a soft and fluffy down quilt that they can snuggle under!

4

I was born in Florida. Snow almost never covers the ground there. I now live in Fairbanks, Alaska, just south of the Arctic Circle. The first sticking snow falls there about mid-October and doesn't melt until late April. Snow lies on the ground for six to seven months of the year. In the Arctic near Barrow, the snow covers the ground nine months of the year. The longer the snow is present, the more important it is for plants, animals, and humans. Here is why:

- Snow reflects sunlight. New snow reflects about 90% of incoming solar radiation. Old snow reflects about 70%. Tundra, in contrast, reflects less than 10%, about one-seventh as much. Once snow covers the landscape, it reflects much of the available sunlight (and there isn't much) back to space, ensuring that air temperatures will stay low.

- Snow insulates the ground. Because snow contains so much air (50% to 80% of a snowpack is air), it is as good an insulator as a down parka or feather quilt. The bitterly cold temperatures experienced by Alaska in the winter would chill the ground much more if it was not protected by the snow. In the North, if the snow doesn't come until after the first cold snap, buried water pipes and septic systems freeze for lack of snow insulation. Cabins that hold snow on the roof are warmer than

those that do not, and without the snow cover, there would be much more, and thicker, permafrost throughout the Arctic than is currently the case.

- Snow melt produces peak river runoff. For most arctic rivers and streams, the peak discharge each year comes when the snow melts. This runoff fills lakes and ponds, floods wetlands, and recharges the groundwater. The Arctic is a dry place, almost a desert, but the snow melt (along with the impermeable permafrost table) ensures that there will be ample water on the landscape each summer. Arctic wetlands attract millions of birds and breed billions of mosquitoes each year, so to them *apun* is also important.

- Snow alters the way humans and animals travel. Alaskans use boats and four-wheelers to get around in the summer. In winter they use snowshoes, skis, snowmobiles, and dogsleds. They change the wheels on their airplanes to skis so they can land on snow. Arctic animals have adapted to the snow too. Moose have long legs and can wade through deep snow. Caribou have wide hooves that spread out when they walk and which help support them on the snow. Lemmings are good diggers and live in snow tunnels during the long winter. Grizzly bears plug up the mouths of their dens with snow so they stay warmer. Even plants have adapted to the snow. Spruce trees have downward sloping branches that are designed to shed snow before they break, while many shrubs are supple so they can lie down under a snow load.

Because of these snow attributes, the snow cover influences directly and indirectly almost every aspect of the arctic natural environment. The effect is different on land and on sea ice but equally important in both domains. Humans, animals, and plants all have had to adapt to snow in order to survive. In the Arctic, snow and life are inextricably connected. The next four illustrations from *Apun: The Arctic Snow* provide examples of the importance of snow cover to life.

Snow and Lemmings

Lemmings live in tunnels at the base of the snow. This keeps them safe from most predators, who can't fit in the tunnels. But a weasel can. The poor lemmings are in mortal danger when a weasel gets into the tunnels. Most of the time, however, the tunnels provide a warm, moist, and safe home.

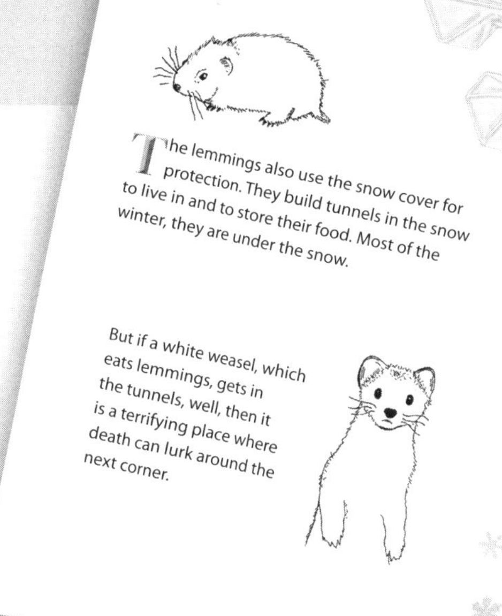
The lemmings also use the snow cover for protection. They build tunnels in the snow to live in and to store their food. Most of the winter, they are under the snow.

But if a white weasel, which eats lemmings, gets in the tunnels, well, then it is a terrifying place where death can lurk around the next corner.

There are two types of lemmings near Barrow: collared (Figure 5) and brown lemmings. Lemmings spend the winter living in tunnels at the base of the snow. This offers them three distinct advantages. First, it is warmer under the snow than above it. In their snug tunnels the lemmings do not have to eat as much to stay warm. Second, the tunnels are probably close to 100% humidity. This is more important than it might seem. In a frozen world, liquid water for drinking is precious. To melt snow for a drink takes significant energy, a high price to pay for an animal that is struggling to survive. The high humidity in the tunnels keeps the lemmings from becoming dehydrated, which would otherwise happen quickly in the cold arctic air.

Third, the snow hides and protects the lemmings from predators like foxes and snowy owls. It does not protect them from weasels (Figure 6), which are small enough to fit in the tunnels. It must be terrifying to be a lemming in a snow tunnel and hear a weasel coming.

Arctic foxes (Figure 7) have developed a different method of getting at the lemmings

Figure 5: A collared lemming. (Photo by Daisy Gilardini/AlaskaStock.com.)

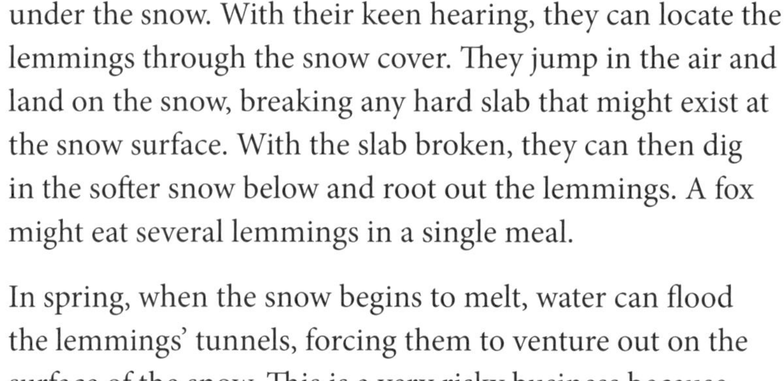

under the snow. With their keen hearing, they can locate the lemmings through the snow cover. They jump in the air and land on the snow, breaking any hard slab that might exist at the snow surface. With the slab broken, they can then dig in the softer snow below and root out the lemmings. A fox might eat several lemmings in a single meal.

In spring, when the snow begins to melt, water can flood the lemmings' tunnels, forcing them to venture out on the surface of the snow. This is a very risky business because birds of prey are waiting to eat the lemmings. Near Barrow, two birds that prey on the lemming are the snowy owl (Figure 8), and the parasitic jaeger (Figure 9). In spring, dozens of jaegers can be seen hovering above the snow, peering down, just waiting for an unsuspecting lemming to venture out from its flooded snow tunnels.

Counterclockwise:

Figure 6: A long-tailed weasel in winter coat. (Photo by Tom Soucek/AlaskaStock.com.)

Figure 7: An arctic fox jumping in the air in order to break the snow crust and get at the lemmings below. (Photo by Gary Schultz/AlaskaStock.com.)

Figure 8: A snowy owl. (Photo by Robert Angell, Alaska Division of Tourism.)

Figure 9: A jaeger. (Photo by Kit Kovacs/Norwegian Polar Institute.)

Snow and Caribou

Perhaps no other animal is more closely associated with *apun* than the caribou. Wonderfully adapted to arctic conditions, no stranger to cold and blizzard, the vast caribou herds thrive as long as the snow stays thin enough, and soft enough, to allow them to dig through it to get at the lichen and plants beneath.

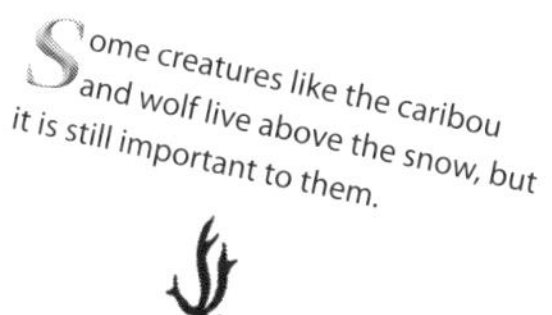

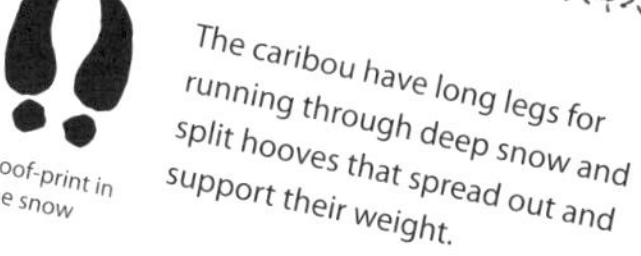

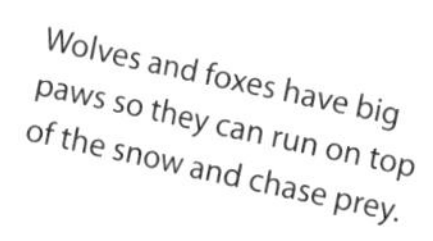

Arctic animals have adapted to living in the snow. Prey species like hare and ptarmigan change color (to white) in the fall so that they are harder to spot (camouflage). Predators like polar bears are also white for the same reason. Seals make sure that the snow cover hides their breathing holes and their pupping dens so that the polar bears cannot find them. Wolverine have developed powerful legs and paws so that they can dig through the snow to capture ground squirrels and lemmings.

The feet of many arctic animals are particularly adapted to traveling on the snow (Figure 10).

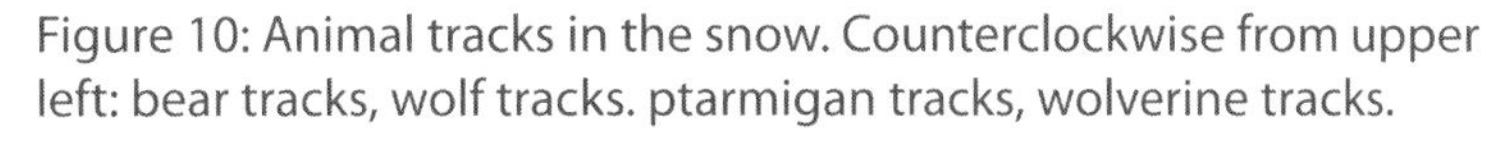

Figure 10: Animal tracks in the snow. Counterclockwise from upper left: bear tracks, wolf tracks. ptarmigan tracks, wolverine tracks.

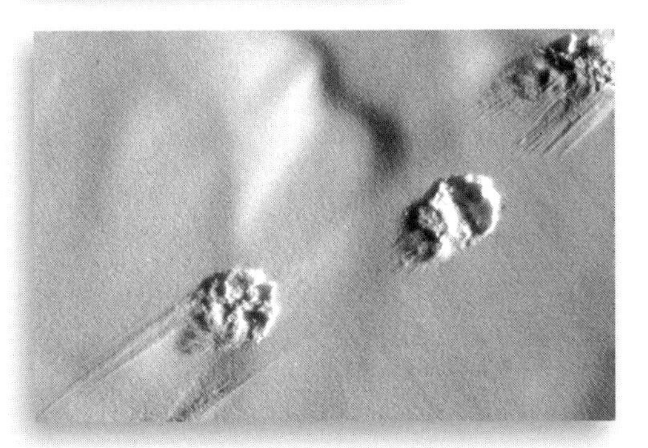

Snow and Shrubs

A willow has been covered by a snowdrift caused when its branches trapped blowing snow. Being buried is good. Once buried, the willow is protected from abrasion and dessication due to the cold, dry wind and abrasive snow particles. It is true that shrubs and other plants grow in summer, but what happens in the winter matters too, making *apun* equally important to flora as well as fauna.

Snow
and
Life

Even the plants need the snow! The snow keeps their roots warm and protects their branches from drying out in the harsh wind. It protects the branches from being "sanded" to death by blowing snow, which in Iñupiaq is *natigvik*. Just like the lemmings, some plants like being under the snow.

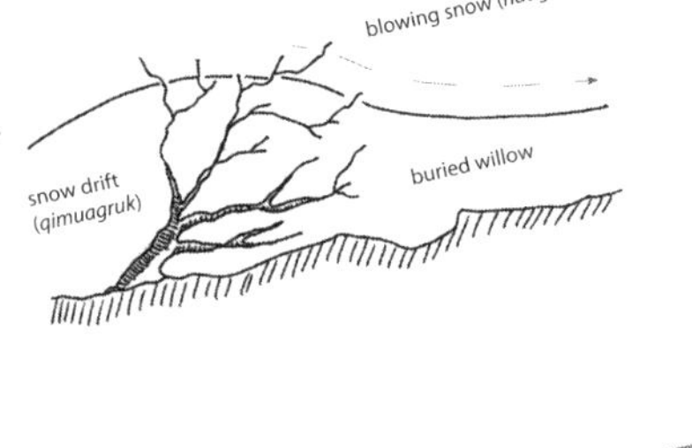

7

Snow bed communities are groups of specialized plants that can tolerate the short growing season (sometimes just a month per year) and the extremely moist conditions imposed by late-lying snow drifts. There is a similarly strong connection between snow and tundra shrubs. This picture (Figure 11), taken near the Ayiyak River, shows the four major types of tundra shrubs (willow, dwarf birch, Labrador tea, and alder) found in northern Alaska. Despite the presence of head-high alder (the largest of the shrubs in this location) the area is still tundra. The particularly tall shrubs in this location may owe their existence to favorable soil conditions produced by the winter snow cover. When winter brings snow and wind, these shrubs will have a marked impact on the snow distribution, which in turn will impact the growing conditions for the shrubs, forming a positive feedback loop.

Figure 11: Tundra near the Ayiyak River in northwest Alaska. Four types of shrubs are present: willow, dwarf birch, alder, and Labrador tea.

In the winter, the shrubs trap windblown snow (Figure 12). This results in deeper snow where there are larger shrubs. The thicker insulating blanket of snow keeps the ground warmer in shrub-rich zones, which promotes more vigorous activity by soil microbes. The more active microbes produce more nutrients, which in turn stimulates the growth of even more shrubs. The cycle feeds back on itself in a positive way to promote shrub growth, which appears to be a contributing factor in the increasing amount of tundra shrubs associated with climate change.

Figure 12: Some large shrubs (willows) have trapped snow that was being blown by the wind (from right to left), creating a drift among and downwind of the most dense branches.

The bigger shrubs also contribute to earlier snowmelt because they poke up through the snow. The snow is white and reflects solar radiation well, but the shrub branches are dark and tend to absorb the radiation (Figure 13). Small melt cavities and chimneys form in the snow around the dark branches as spring progresses. Millions, or perhaps billions, of these little shrub-induced melt accelerators add to the overall melt rate and lead to an earlier spring on the arctic tundra. It has been documented that spring now comes seven to ten days earlier in northern Alaska than it did thirty years ago, and the shrubs are probably contributing to some of this change.

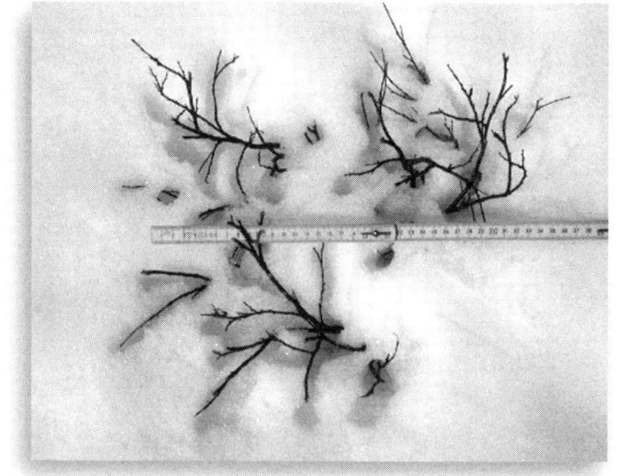

Figure 13: Dark shrub branches sticking up through the snow absorb solar energy, creating cavities and little chimneys. This contributes to accelerating the melt rate.

Snowmobiles

Two people from Barrow are out on the tundra on snowmobiles, the main vehicle used for travel in the winter. In Alaska, people call them snowmachines. The snowmobiles have tracks for pushing against the snow and skis for steering to take advantage of the fact that sliding is easy on the snow. One fellow is about to "get air" as his machine launches from a snow ramp. Much larger tracked vehicles are also used throughout the Arctic for working in winter.

Apun, the arctic snow cover, means good traveling for people—and fun too!

Thousands of years ago, humans discovered that it was easier to slide something along on top of the snow than to drag it over bare ground. This led to the invention of the ski. There are rock drawings of skis dating back almost 5,000 years, and skis preserved in European bogs that are even older. By the twelfth century (Figure 14) the ski was already a key part of Nordic culture. As this famous painting shows, the two Vikings, who are carrying two-year-old Prince Haakon from Lillehammer to safety in Osterdalen, are wearing quite elegant skis. This historic event took place in 1206!

In Canada and the United States, indigenous people developed toboggans and snowshoes. A toboggan is a narrow plank sled (Figure 15) designed to be towed behind a person walking on snowshoes (Figure 16). Written references to toboggans date back to the early 1700s, suggesting an ancient origin. The narrow toboggan was primarily designed

Figure 14: Two Viking warriors on skis rescue the child-prince Haakon in the year 1206. (Courtesy of the Mammoth Ski Museum.)

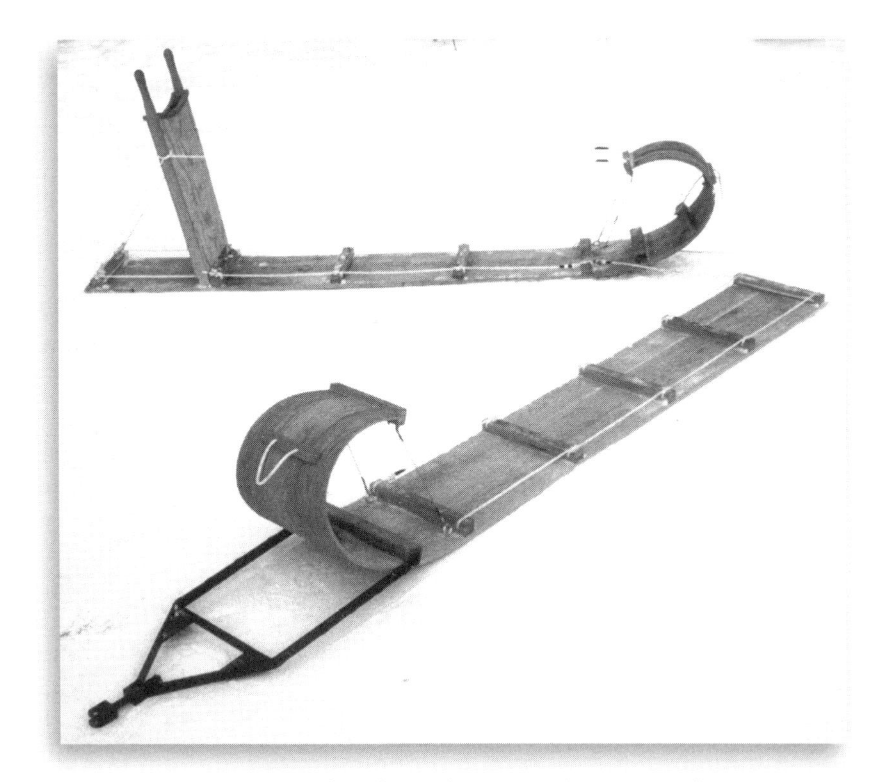

Figure 15: Two replicas of traditional towing toboggans. (Photo provided by John Harren of northern toboggan and sled.)

Figure 16: A set of huge snowshoes designed for the deep, soft snow of the taiga forest. (Photo by Jon Holmgren.)

for use in the taiga forest where the snow was soft and deep and the trails between the trees were narrow. Big snowshoes to provide adequate floatation were also needed for use in this type of arctic snow cover (Figure 16).

For the tundra, with its harder packed snow and wide-open spaces (no need to stay on narrow trails), runner sleds, much wider than a toboggan, were developed. For traveling over the snow on the sea ice, sleds were built with thick solid runners that could take the hard impact of hitting ice. Today, there are dozens of types of sleds in use in the Arctic (Figure 17). The designs and materials of the most recent sleds can be traditional, but more often both the designs and the materials are modern.

The need to adapt to a snowy world has led to the development of not only many types of sleds, but other technologies as well, including aircraft on skis (Figure 18) and powerful snowblowers (Figure 19) that can cut through snow-drifts more than 10 feet (3 m) deep.

Figure 17: Sleds in the Arctic. (A) A dog-mushing sprint sled (Old Crow, Yukon Territory). (B) A modern *komatik* (sea-ice sled) for hauling a boat to a lead in the sea ice (Barrow, Alaska). (C) A more traditional *komatik* (Kugluktuk, Northwest Territory). (D) A long-distance dog-mushing sled made of plastic and aluminum (Old Crow, Yukon Territory). (E) A traditional toboggan (Ft. McPherson, Yukon Territory). (F) A heated hut-sled used as a mobile kitchen (Barrow, Alaska). (G) A specialized heated sled for scientific work (Porcupine River, Alaska). (H) A new Koyuk sled of traditional design and materials (Buckland, Alaska).

Figure 18: A Twin Otter aircraft on skis, landing on snow-covered Daring Lake, Northwest Territories, Canada.

Figure 19: A powerful snowblower clearing a drifted road.

Snow Crystals

Condensation Nuclei

The picture shows a salt particle on which water droplets are condensing. Without the salt particle as a base (called a condensation nuclei), the water would not condense until the temperature dropped below –40°F (–40°C). The salt particle has been greatly exaggerated in size to emphasize the role it plays in creating snow crystals. In fact, it is tiny: much smaller than the ice crystal that forms. It is small but very important.

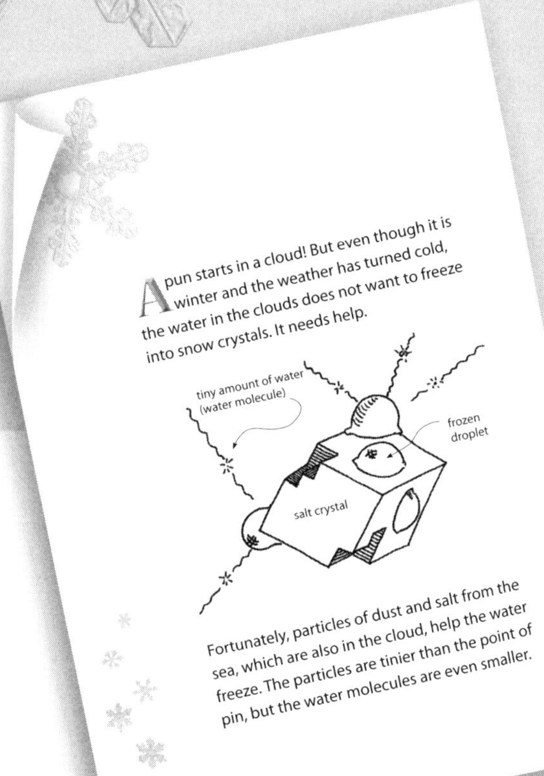

Apun starts in a cloud! But even though it is winter and the weather has turned cold, the water in the clouds does not want to freeze into snow crystals. It needs help.

Fortunately, particles of dust and salt from the sea, which are also in the cloud, help the water freeze. The particles are tinier than the point of pin, but the water molecules are even smaller.

10

The arctic snow cover (*apun*) is built up one snowflake at a time. I once had to estimate how many snow grains were in the snowpack near Fairbanks for a study I was doing. There were about 400 million grains per cubic meter. That is about 11 million grains in a cubic foot. A football field covered by two feet of snow would support more than a trillion snow grains (1,000,000,000,000 grains). Due to metamorphism after deposition, these grains have different shapes and sizes than snowflakes, but without snowflakes to start with, there would be no snow pack, no *apun*.

Surprisingly, growing beautiful and symmetric crystals of ice is not as easy as it might seem. In part, this is because snow is "warm." We tend to think of snow as a cold material, and it is, compared to the temperature of the human body or a boiling kettle of water. But another way to think of snow is that it is rarely at a temperature much lower than its melting point of 32°F or 0°C. Even on a cold day in the Arctic, when the air temperature drops to –40°F (which is also –40°C), the snow is only 72°F (40°C) colder than the temperature at which it melts. In contrast, a bar of iron melts at over 2500°F (1500°C), so at room temperature it is more than a thousand degrees below its melting point. If the iron bar were only 72°F below its melting point, it would have very different properties than iron normally does. It would bend and deform easily (Figure 20). It would not be very strong, and if brought in contact with other pieces of iron it would fuse to them. Driving over an iron bridge behaving this way might be exciting!

In addition to being a "warm" material, snow is made from water molecules that have a strong polarity (electrical charge) that makes them prone to supercool. That means they resist freezing. It is not unusual for tiny droplets of liquid water to still be present in a cloud at –40°F (–40°C). Minute ice crystals (the start of snowflakes) could possibly condense directly from water vapor in a cloud, but the rate would be so slow, it is certainly not the main mechanism that produces the snowfall rates we see on Earth. A much faster mechanism must be at work, and it is. That mechanism is one in which water vapor condenses initially on tiny foreign particles (called condensation nuclei) of sea salt, soot, sulfate, and even plant material. These nuclei speed up the process of forming both raindrops and snowflakes. Sea salt particles, very common near coastal Barrow, Alaska, end up in clouds when the wind whips up ocean spray (Figure 21) or when bubbles burst at the sea surface and spray particles into the air. Soot particles come from forest fires and smoke stacks. Sulfate particles come from volcanic eruptions and phytoplankton. Recently, scientists have even found that there is organic plant material in the atmosphere that is particularly effective at nucleating snowflakes.

I drew the salt crystal in the black-and-white drawing greatly exaggerated in size. I wanted it to look like a salt crystal. In fact, a water droplet formed around a salt nuclei would be 100 times larger than the nuclei itself, and a raindrop formed from a collection of these micro-droplets would be 100 times bigger than the individual droplets, making the raindrop more than 10,000 times larger than the condensation nuclei at its center (Figure 22).

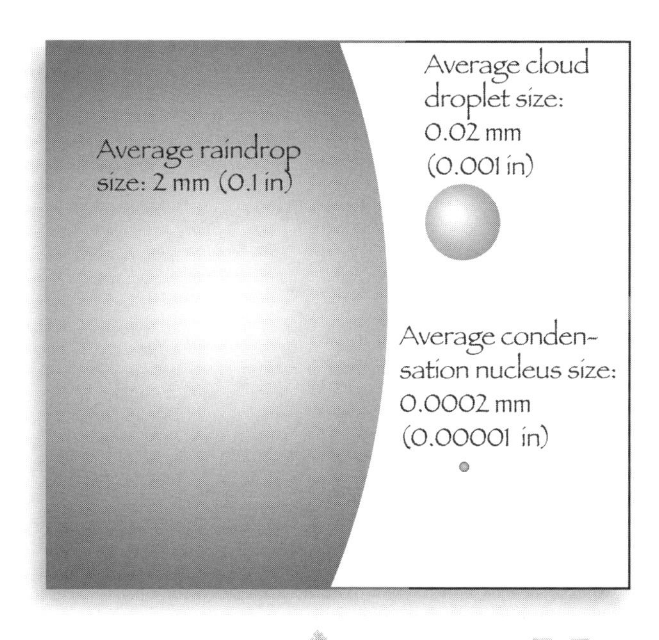

Top to bottom:

Figure 20: A red-hot iron bar behaves differently than cold iron because it is close to its melting temperature. Snow that is close to its melting temperature behaves in a similar way, which is why it metamorphoses and changes so readily. (iStockphoto.com/Kawisign.)

Figure 21: An ocean wave lofts salt spray into the air, a source of cloud condensation nuclei. (iStockPhoto/Hanis.)

Figure 22: The relative sizes of raindrops, cloud droplets, and condensation nuclei.

Diamond Dust

Diamond dust ice crystals in a cloud are tiny but have very regular geometric shapes. They are simple but elegant crystals that produce dramatic displays of light in the arctic sky.

As the cloud droplets freeze around dust and salt particles, tiny ice crystals form. Five crystals could fit on the point of pencil. They are so small and light, they float in the air instead of falling to the ground. They will need to grow a lot before they can pile up on the ground as Apun. As they float, they sparkle in the sun like diamonds.

This is called diamond dust, or in Iñupiaq, *irriqutit*.

If the air temperature in the cloud is well below freezing, then ice crystals rather than water droplets will condense out. If these form near the ground under clear skies, it is called diamond dust. The dust crystals (which are really made of ice) are tiny: 50 to 250 μm (0.05 to 0.25 mm, or 2 to 10 thousandths of an inch). They are so small and light that they settle through the air very slowly, their limited weight balanced by their air resistance. You might say they have a long hang time. As a consequence, the air will be filled with floating crystals that sparkle in the sunlight. This is where the name diamond dust comes from.

These minute crystals tend to take a limited number of forms, although all have hexagonal symmetry (Figure 23). The crystals can be short stubby columns, long thin columns, hexagonal plates, or needle-like crystals called shimizus. The type of crystal that forms is primarily a function of the temperature. Shimizus form on cold days (–40°F), while plates and columns form when it is warmer. Unlike snowflakes, which are larger and have had more time to grow, diamond dust crystals are elegant but simple. Unfortunately, they are so small it is difficult to view them without a good microscope.

A key characteristic of diamond dust is that it forms low in the atmosphere, within a few hundred feet of the ground, and it forms during clear weather, with no clouds in the sky. It can form at night or in the daytime, but it takes its name from its appearance by day: the sky seems to be filled with millions

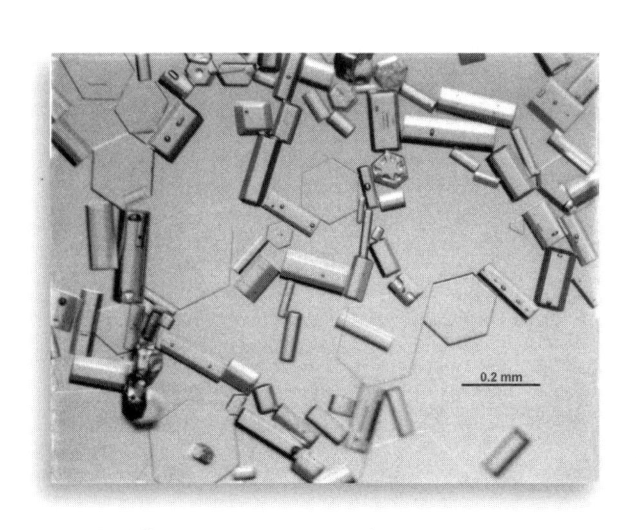

Figure 23: Diamond dust crystals. (Photo by Walt Tape.)

Top to bottom:

Figure 24: Sun dogs, arcs, and halos are caused by diamond dust in the air. (Photo by Walt Tape.)

Figure 25: The orientation of plate and column crystals as they fall through the atmosphere.

Figure 26: The relationship between the light display and the type of diamond dust crystal. (Image by Matthew Sturm and Les Cowley.)

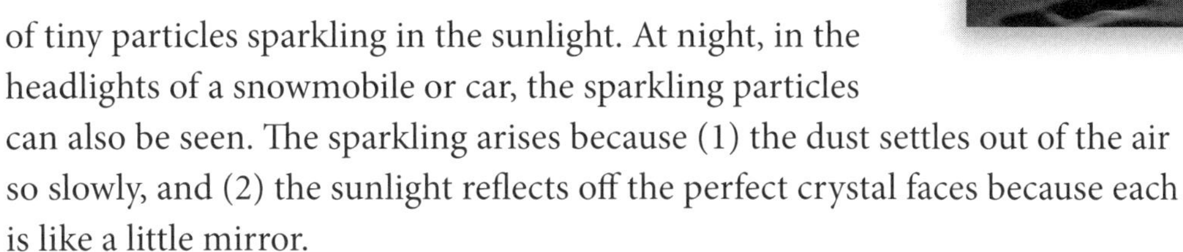

of tiny particles sparkling in the sunlight. At night, in the headlights of a snowmobile or car, the sparkling particles can also be seen. The sparkling arises because (1) the dust settles out of the air so slowly, and (2) the sunlight reflects off the perfect crystal faces because each is like a little mirror.

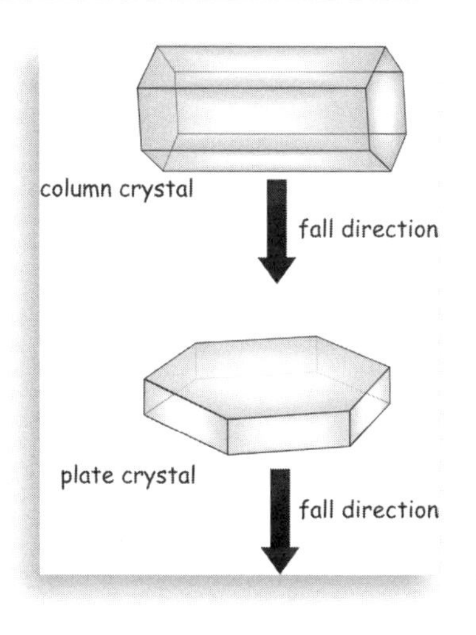

The interaction of diamond dust and sunlight also gives rise to sun pillars, perihelia (sun dogs), halos, and other spectacular lighting phenomena (Figure 24). These phenomena are most often seen in the polar regions, but they can be seen in lower latitudes as well. In fact, small rainbow spots are seen near high cirrus (ice) clouds in the tropics. These too are actually formed by diamond dust.

Scientists understand how the arcs, halos, and sun dogs form. Recall that diamond dust has three basic forms: hexagonal plates, long hexagonal columns, and shimizus (needles). The plates and columns settle through the air in different ways (Figure 25). The plates have the maximum air resistance on their broad hexagonal surfaces, so they settle like leaves, with their hexagonal surfaces facing up and down. The columns have their maximum air resistance on their long sides, so they settle the other way, with their hexagonal surfaces vertical. Within a cloud of diamond dust particles, most of the plates or columns will be lined up within a degree or two of the orientations shown in Figure 25. The result is that the dust makes up a giant prism, composed of millions of individual crystals, all nicely aligned. The prism reflects and refracts sunlight in a predictable way. Various layers of crystals at different heights in the air refract and reflect sunlight to make the beautiful and intricate light displays pictured in Figure 26. The process is not all that different from hanging a crystal prism in a sunny window and having it produce a pattern of rainbow spots on the walls.

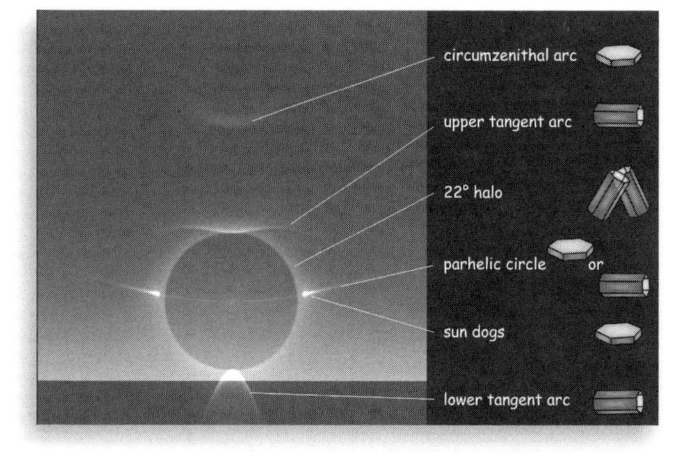

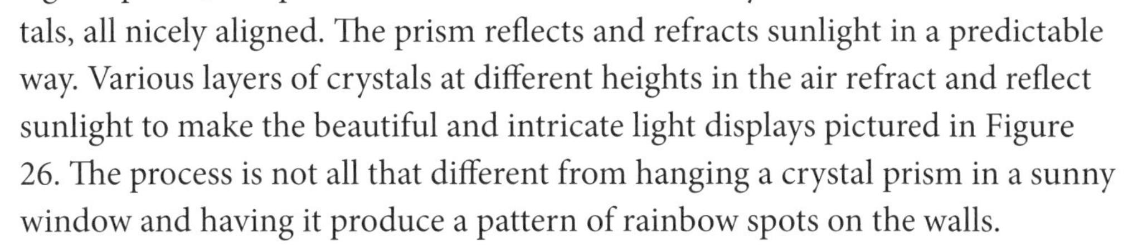

Snowflakes

Changing diamond dust into large snowflakes requires rapid growth. How does that happen and why do snowflakes look like they do?

The clouds over the ocean are moist. The diamond dust is blown up and down in the clouds. The crystals eventually grow into snowflakes (*qannik*) heavy enough to fall out the bottom of the cloud. The falling crystals have perfect hexagonal shapes and funny names.

sector plate

1 mm

stellar dendrite

12

Snowflakes can be one hundred times larger than diamond dust crystals, and they are always more ornate and complex. Like diamond dust, snowflakes have a basic hexagonal symmetry, but they can display such a bewildering range of shapes and forms that four questions come to mind:

1. How do they grow so large?
2. Why are they (mostly) symmetrical?
3. Are no two snowflakes identical?
4. Why do so many different forms of snowflakes develop?

The high growth rate of snowflakes in a cloud is still something of an unsolved problem in physics, but a key factor is supersaturation. The air in a cloud from which snowflakes are falling is actually holding more water vapor than it should. The least little cooling, or the presence of any favorable condensation nuclei, and that water vapor starts to condense instantly. A familiar example of this sort of effect is the rapid formation of a fog bank in the evening when the temperature cools. Initially simple snow crystals without arms or branches form and grow, producing hexagonal plates like the one in Figure 27.

Slow fall rates and updrafts in the cloud keep these simple unbranched plates from dropping out of the cloud, but eventually they grow heavy enough that they begin to fall. They travel downward through cloud layers that have a different temperature and humidity than where the flakes originated. The crystals continue to grow. Initially their small size favors the growth of edges (the long

Figure 27: A simple, unbranched snowflake. (Photo by Ken Libbrecht.)

Left to right:

Figure 28: A snowflake with stubby branches. (Photo by Ken Libbrecht.)

Figure 29: A stellar dendrite snowflake: most people's idea of a classic snowflake. (Photo by Ken Libbrecht.)

straight sides) rather than points, but eventually the crystals become large enough that point growth starts to take over. Stubby branches develop (Figure 28).

Because the branches stick out into the supersaturated air farther than the edges of the crystal, they now grow even faster. The faster they grow, the more they stick out, a self-reinforcing process. The result is that stellar dendrites (Figure 29) are formed.

The great symmetry of snowflakes is initially due to the structure of the water molecule, which results in beautifully symmetric diamond dust and small unbranched snowflakes. As the snowflakes grow larger, they tumble and rotate as they fall. Each branch or arm is exposed to much the same growth conditions as the other five arms, so they all tend to grow the same. In reality, non-symmetrical snowflakes are quite common, but as they are not as photogenic as symmetrical ones, they tend to be photographed less.

The long-running argument as to whether any two snowflakes are identical is really more semantic than real. At the molecular level, no two snowflakes are likely to ever be the same. More practically, the question might be whether any two snowflakes look the same. The answer to this question is certainly yes. Particularly for simple, unbranched crystals and diamond dust, it is highly likely that two snowflakes are so similar as to look identical to a human observer.

The two snowflakes shown in the black-and-white *Apun* illustration are a stellar dendrite (left) and a sector plate (right) with stubby branches, which are the snowflake types that most people think of when they imagine a "classic" snowflake. However, there are many other forms, some of which are quite common in the Arctic. Magono and Lee, two snow scientists, developed a classification system that recognizes eighty different forms, the bulk of which are not "classic" in shape. Here (Figure 30), thirty-five different forms are shown.

Figure 30: Thirty-five types of snowflakes based on the classification of Magono and Lee (Meteorological Classification of Natural Snow Crystals, *Journal of the Faculty of Science*). (Redrawn by Ken Libbrecht. Used with permission.)

The reason that there are so many different crystal forms lies in the journey each snowflake makes from the time it forms high in a cloud until it lands on the ground. The crystal form is controlled by the temperature and supersaturation of the various layers. Each combination imposes on the crystal different shapes. By the time the crystal lands on Earth, it is the end result of a complex and unique history of growth and travel. Ukichiro Nakaya (1900–1962), a Japanese physicist who conducted the first systematic studies of snow crystal formation, called snowflakes "messengers from heaven." By carefully growing snow crystals in a cold chamber, he was able to determine how crystal habit was controlled by temperature and supersaturation (Figure 31). Based on this knowledge, he could deduce the temperature and humidity in the cloud layers by merely observing the shape and form of the snowflakes that fell.

Nakaya published his results in 1954 in a book entitled *Snow Crystals: Natural and Artificial*. The book is a wonderful mixture of art and science. Today Nakaya is honored for his work in Japan with a museum in his hometown (http://www.city.kaga.ishikawa.jp/yuki/index-e.html) and is on a Japanese postage stamp.

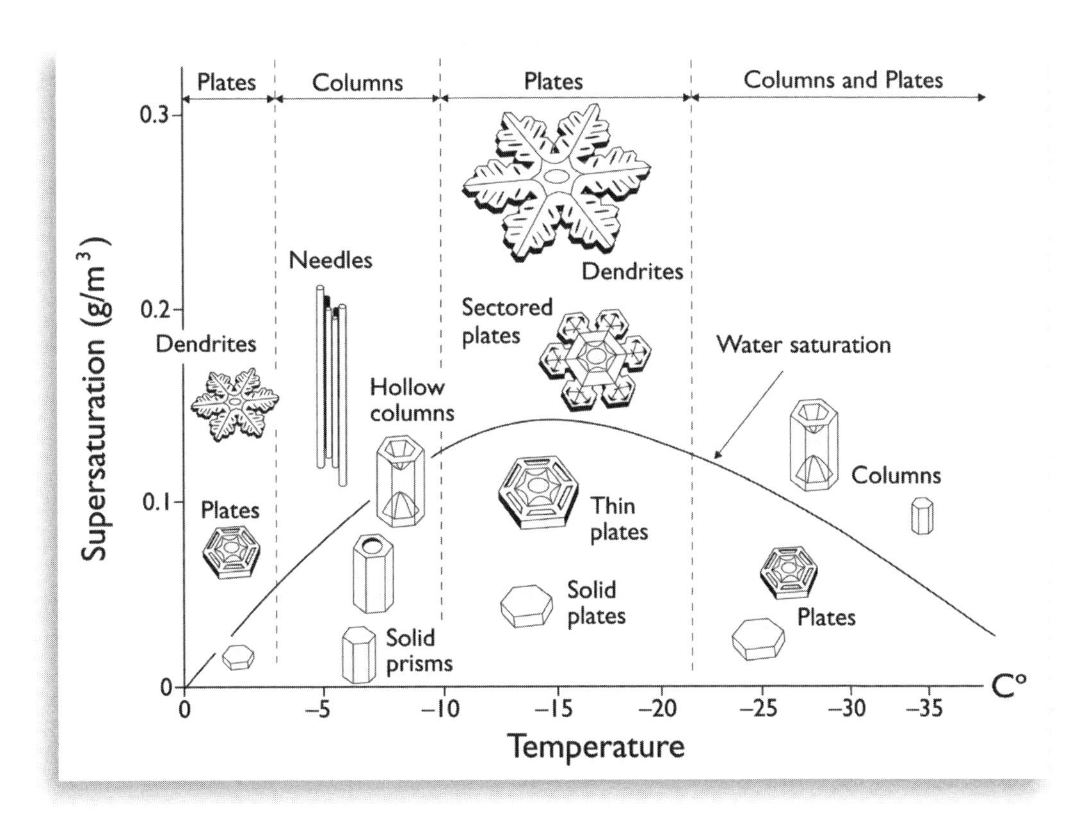

Figure 31: Snowflake morphology is controlled by temperature and supersaturation. This morphology diagram comes from Ken Libbrecht's *Field Guide to Snowflakes* (Voyager Press, 2006). The original version of this diagram was developed by Ukichiro Nakaya and was called a Nakaya diagram.

Capped Columns

Strange crystals, barely recognizable as snowflakes, fall regularly in Barrow.

Sno
Crysta

capped column

capped column

bullet rosette

1 mm

Some snowflakes are hexagonal but don't look anything like stellar dendrites or sector plates.

13

Because Barrow is at high latitude, it experiences low winter temperatures, but because it is near the ocean where there are open leads in the sea ice, it often has high humidity. These two conditions, which are not that common, are perfect for the development of capped column and bullet rosette snow crystals (Figure 32). We routinely see these strange snowflakes when we make snow observations in Barrow. They can be seen at lower latitudes, but not as frequently.

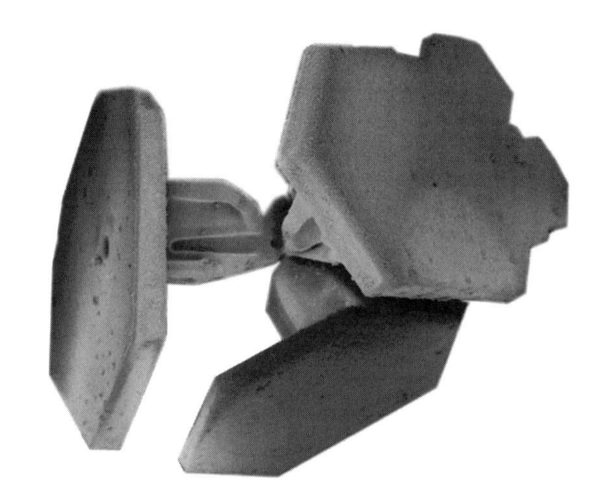

Figure 32: A scanning electron microscope image of a bullet rosette with very large column caps. (Image by Eric Erbe.)

Water Molecules

The structure of the water molecule creates the hexagonal symmetry seen in ice crystals, diamond dust, and all the forms of snowflakes.

The symmetry and crystalline perfection of diamond dust and snowflakes arises from the geometry of the water molecule, which consists of one oxygen atom (O) and two hydrogen atoms (H_2). Chemists represent that configuration with the formula H_2O. The actual structure in ice, however, is a little more complicated. Each oxygen atom forms the center of a tetrahedron (like a three-sided pyramid) with the hydrogen atoms making up the four tetrahedral corners (Figure 33). Each hydrogen atom is shared between two oxygen atoms (that is why the formula is H_2O rather than H_4O), making a very strong bond between each tetrahedral group. The whole mass, when viewed from above or below, has a hexagonal outline (Figure 34) that is mirrored in the crystal forms. In some ways it is like building with Lego blocks: many shapes can be built, but the shapes invariably reflect something of the fundamental shape of the blocks themselves, which in this case are hexagons.

The perfect shapes of these crystals happens because they are made of water molecules which look like this:

hydrogen atom

oxygen atom

When several water molecules are linked together, it looks like this:

Can you see why a snowflake would have six arms and a hexagonal shape?

14

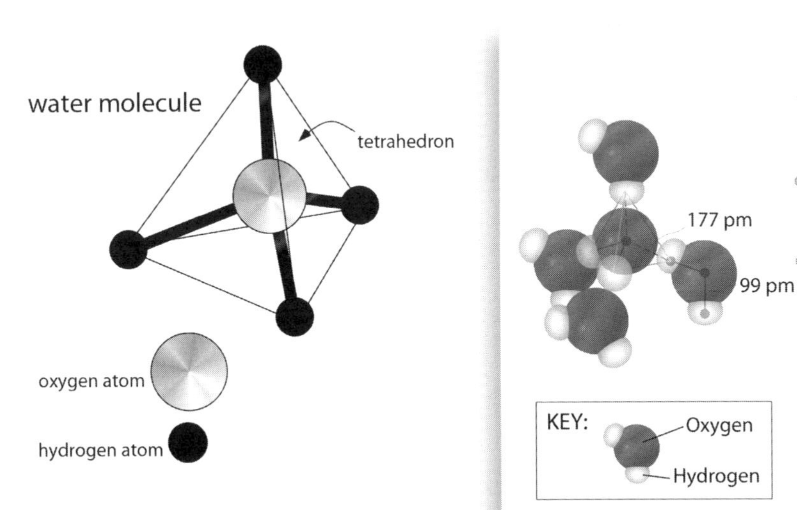

Left to right:

Figure 33: The water molecule has a tetrahedral shape, with the oxygen atom in the center and the hydrogen atoms forming the four vertices.

Figure 34: When the oxygen-hydrogen tetrahedra are connected together in a larger matrix, beautiful hexagonal symmetry is the result. The lengths of the bonds are shown (pm=picometers, or 0.000000001 mm).

Changing Snow

October Snow

The picture shows an early snowfall that has mostly melted, with the remaining snow on the north side of the tundra tussocks where it was shielded from the sun. This would be typical in October, an "in-between" time for traveling in the North. There is not enough snow for sleds, but perhaps too much snow for good walking. The fact that there are two words for melting snow, one for fast (*auksiqlaq*) and one for regular (*auktaaq*), indicates how important this transition period is to hunter-travelers like the Iñupiat. It may seem strange, but melting is one form of snow metamorphism.

The first thing that happens in October is that fluffy new snow (*nutagaq*) falls from the sky and soon covers the tundra. The tundra is likely to still be warm from the summer, so this first snow often melts immediately (*auktuq*), leaving the tundra a patchwork of bare tundra and wet snow.

augniqsraq (a bare patch from which snow has melted)

16

Before a snow cover can develop on the tundra, the ground has to freeze. All through the summer, the tundra warms up and the ground thaws from the top down. The dark color of the tundra helps it warm up because it absorbs sunlight. On a sunny day, the tundra surface can be 80°F (27°C). The thawing will penetrate downward 24 to 36 inches (60 to 90 cm) by the end of the summer. The layer that thaws and freezes each year is called the active layer. Below the active layer the ground stays permanently frozen (permafrost, Figure 35), which also means water cannot infiltrate downward. Consequently, the thawed soil is usually wet, often completely saturated. So when freeze-up comes in September, there is a lot of heat stored in the soil and the unfrozen water. If it snows before the weather has turned cold, the first snow will hit the tundra and melt on contact. Eventually, sometimes as late as November, the soil surface will finally freeze, and then the snow will stick (see Figure 4 on page 6).

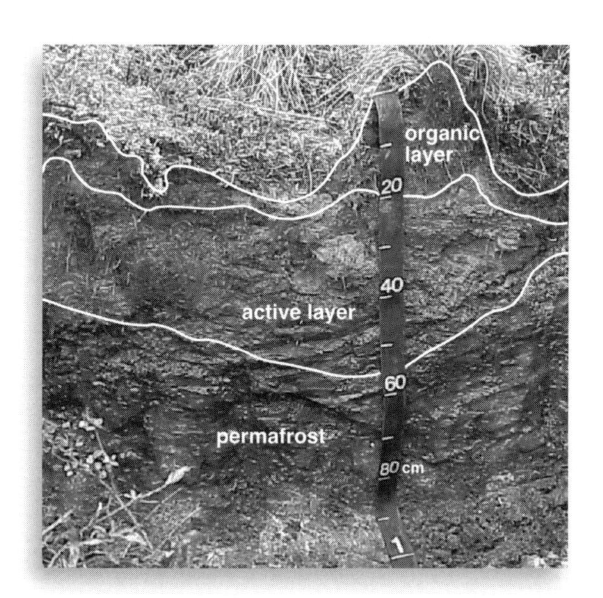

Figure 35: A soil pit showing the active layer underlain by permafrost.

Snow Metamorphism

By November, there is a continuous blanket of snow covering the land. The snow cover seems to be white, unchanging, and static, but in fact unseen, down in the snow, changes are taking place. Wind, heat, cold: all of these are working to alter the original snowflakes that fell into new shapes and forms. The circles at the bottom of the *Apun* illustration show how a new snowflake might change over the period of a week under these metamorphic forces. First the snowflake will break apart and start to lose its arms and points. Next, these smaller, rounded grains may start to grow sharp, geometric facets (like a cut gemstone). Finally, they will grow into cup-shaped crystals with ornate features.

Changing Snow

Soon, however, it gets colder and the snow no longer melts. The tundra gets completely covered and looks white. But unseen, down in the snow and at the snow surface, the snow crystals continue to change. This process has a fancy name: "snow metamorphism." The snowflakes change into something quite different by the end of the winter.

Once the snowflakes have piled up on the ground (or sea ice), changes start to take place, driven by weather conditions, solar heating, and gravity. Three basic forces alter the snow: wind, temperature gradients, and heat. The three forces are known as metamorphic forces, and they can produce large changes in the characteristics of the *apun*. Figure 36 shows the metamorphic pathways that crystals can follow.

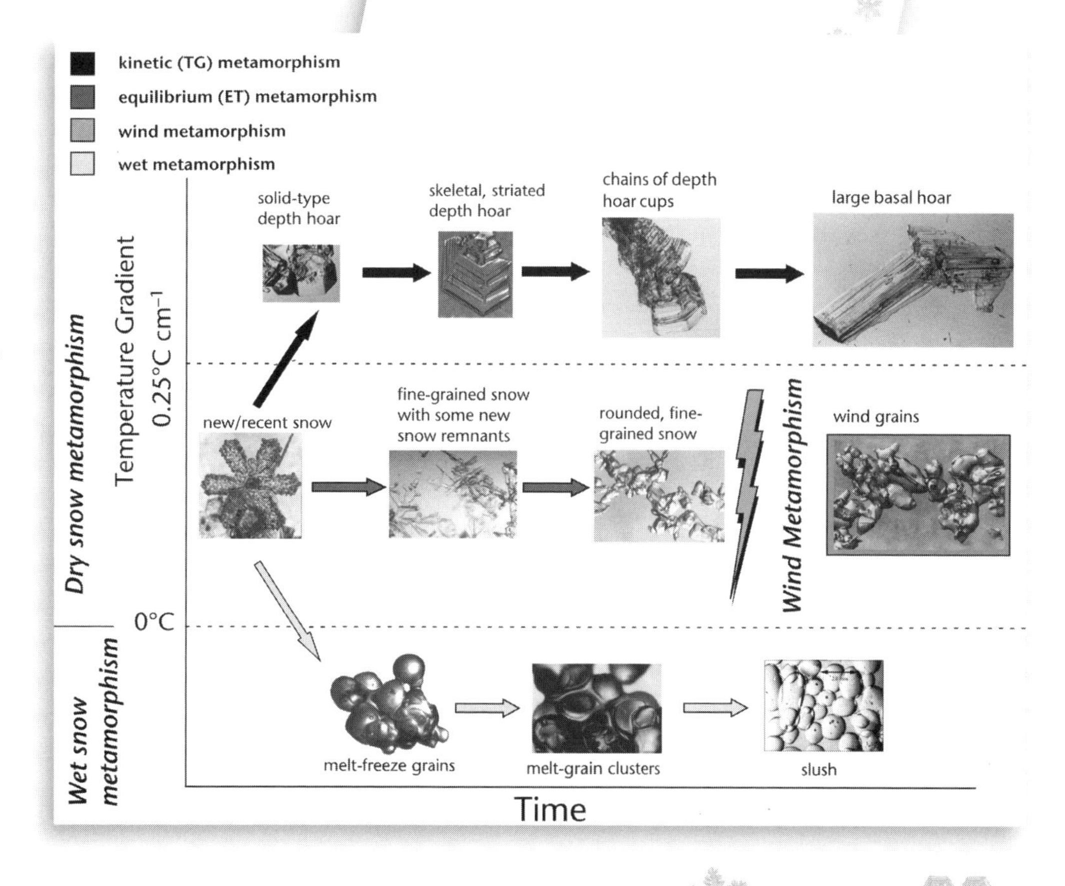

Figure 36: Metamorphic pathways for snow on the ground.

Three Metamorphic Forces

While there are an infinite number of forms that snow crystals can take, there are really only three forces that shape and control the processes that change the snow: wind, differences in temperature from the bottom to the top of the snow pack (called temperature gradients), and heat.

There are three forces that alter and change snow crystals:

Wind (*anuġi*)

Temperature Gradient (*qanuġinnina siḷam*)

Heat (*uunnaq*)

18

As soon as the snow sticks and covers the ground it begins to change or metamorphose. Three forces drive this metamorphism and are critical in creating the arctic snowpack:

1. The wind can blow the snow about, pulverize the grains, yet ultimately make some of the strongest layers in the snowpack through a strange process called sintering.
2. Differences in temperature from one part of the snowpack to another (called temperature gradients) can drive water vapor to move from one grain to another through sublimation and condensation. The *Apun* drawing suggests these differences in temperature by juxtaposing a campfire (warm) and an icicle (cold).
3. The sun, or above-freezing temperatures, can melt the snow, producing smooth rounded grains, ice layers and lenses, and icicles in the snowpack through a process called percolation.

The three forces produce very different types of snow grains and therefore very different *apun* properties. First we will examine what the wind does to the snow, followed by temperature gradients, and then finally heating.

Wind

One powerful force that acts on the snow is wind, the harsh and ever-present arctic wind. The wind does its work by using a trick to get the snow moving and blowing about. It is called saltation, and it is a little like a leapfrog game for grains of snow.

Wind

Changing Snow

Soon, however, the tundra cools off, and the arctic wind (*anuġi*) starts to blow. It is the time of the blizzards. The wind blows and tumbles the newly fallen snowflakes, breaking them into pieces. If you were to watch closely when the wind is driving the snow, it would look like thousands of tiny white grasshoppers or *miñŋuq* (small arctic beetles) jumping and bounding along, crashing into each other, and jumping again.

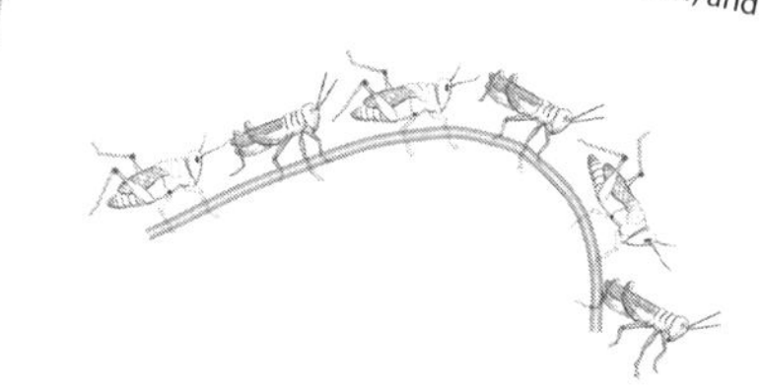

19

How the arctic wind can howl! With no trees to block the wind, and the tundra stretching away seemingly for endless miles, sometimes Barrow feels like the home of the wild wind. I recall after five days of a continuous windstorm, one of my colleagues asked me plaintively, "Why doesn't it run out of wind?"

What makes the windy Arctic seem less hospitable than equally windy southern climates is blowing snow. You have to experience a blizzard to understand the nerve-wracking dimension it adds to windy weather. The visibility is bad, pulverized snow gets into your clothing and equipment, and anything on the ground gets drifted over and buried in minutes (Figure 37).

Figure 37: Ken Tape and April Cheuvront shelter behind a snowmobile during a ground blizzard in northern Alaska. Their clothes are covered by finely pulverized drift snow.

Snow Saltation

Anyone who has seen or played pool knows about collision and rebound. That, plus the fact that the wind is stronger higher above the snow surface, are the ingredients for saltation.

Of course, it isn't really white grasshoppers or beetles: it's jumping grains of snow, and there is a trick to how the wind moves these grains. It is called saltation.

One snow grain will be driven by the wind into second grain on the ground.

This will bounce the second grain up into the air!

wind strength

The wind is stronger higher up, so the second grain goes flying down the wind, only to crash into a third grain, and so on.

20

Snow on the ground will start to move when the wind hits about 10 miles per hour (5 meters per second). Ordinarily, this would not be sufficient to move most snow grains, but a leapfrog-like process called saltation allows the wind to get snow grains moving more easily. Saltation takes place because the wind a few inches above the snow surface is much faster than the wind right at the snow surface. The process works like this: a grain of snow, perhaps a bit lighter than its neighbors, or sticking up a bit more, is caught by the wind and starts bounding along. As the grain moves downwind, it strikes another grain. The momentum of the first grain is just enough to pop the second grain up a little above the snow surface, where it is now caught by the faster wind. It starts moving downwind. Sometimes the first grain will bounce two grains into the air. These now blow downwind too, and when they fall back to the snow surface, they eject still more grains. Soon all of the surface snow is in motion. The grains follow perfect parabolic trajectories. A Japanese scientist, D. Kobayashi, filmed these trajectories in the 1970s (Figure 38) by mounting a camera at the snow surface in a wind tunnel filled with snow.

Figure 38: Saltating snow grains in a wind tunnel in Japan. (Photo by D. Kobayashi.)

Blizzards

The Iñupiat man in the blizzard is a seal hunter. His staff is for hooking seals before they sink. He is smiling in the first panel because the snow is saltating only about ankle-high. The visibility is good and the sun is shining. As the wind gets stronger (shown by the black arrow), the drifting snow rises to his waist and then to his head. As the wind increases in force, the hunter has to lean into the wind, and the visibility gets worse and worse, with the blowing snow eventually obscuring the sun and full whiteout conditions developing. Three different Iñupiaq words capture these essential differences.

Changing Snow

There are many words in Iñupiaq for blowing snow.

WIND

ankle-high drift (*natiqlit*)

knee-high drift (*natigvik*)

full-on blizzard (*agniqsuq*)

21

At 10 miles per hour wind (5 meters per second), snow starts to saltate (Figures 38 and 39). As the wind gets stronger, the grains leap higher and are blown farther. At about 20 miles per hour (10 meters per second), the saltating grains will begin to break apart when they impact other grains, creating smaller and lighter grains that can be lofted by the wind. As long as the wind continues to blow, these smaller grains don't come down. They float and swirl on the turbulent air, in the case of very high winds, rising more than 100 feet above the ground. These high-flying grains are called the

Figure 39: The drifting snow layer is almost up to Glen Liston's waist, suggesting the wind speed is approaching 20 mph. The layer is mostly saltating (bouncing and rebounding) snow, rather than a suspended load. We can tell this because the visibility is still pretty good. (Photo by Henry Huntington.)

suspended load. Once the wind is strong enough to transport both saltating and suspended grains, the visibility drops dramatically. This produces the full-on blizzard and whiteout conditions most people think of when they hear about arctic blizzards.

During one of our arctic expeditions we got caught in a severe blizzard while traversing a long and exposed ridge. We had to get off the ridge and find shelter. Thankfully, we had a good map and a GPS, which allowed us to find and navigate our way off the ridge and down into a more sheltered valley. We got the tents up, and finally we could relax. Looking across the valley, I couldn't see more than 100 feet, yet looking straight up, I could see the sun shining down on us, albeit looking a bit hazy. The layer of drifting snow could not have been more than 50 feet thick. It was not snowing, yet our clothing and equipment were rapidly being covered in a powdery substance that looked like snow but had the consistency of flour. It clung to surfaces in ways normal snow never does. This material was the suspended snow load falling out of the sky, pulverized by the wind so that it was as fine (and as small) as diamond dust. Its odd clingy behavior was due to its unusually small particle size.

Sintering

Why do people build igloos but not sandloos? The wind blows sand and snow into drifts and dunes, but only the snow hardens up. Why? The answer is a process called sintering. It happens for snow and some metals.

When the wind stops blowing, the broken snow grains become "glued" together in a super-strong mass. The glue joints are called bonds, and the thicker they are, the stronger the snow. This process is called sintering. Some snow is so hard it can only be cut with a saw. Depending on how hard the snow is, it might be called *aniu* (softer), *aniuvak* (harder), *nuturuk* (firm, good for making a snow house), or *siḷḷiqsruq* (super-hard, often icy).

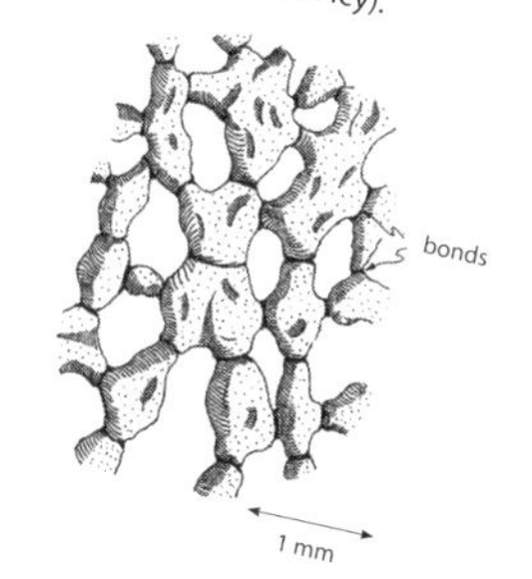

22

How can a cornice made of snow curl like a wave without falling apart or breaking? The answer is a process called sintering. It is a process unfamiliar to most people, yet many of the parts in a cell phone are produced through sintering. Pulverized metal is packed into a mold, heated, but not hot enough to melt it, then allowed to cool. When removed from the mold, the metal will have bonded into a solid mass. This happens because anywhere two metal particles touch each other, the particles exchange molecules and bond into a single mass. Snow does the same thing. Unlike metal sintering, however, heating the snow is unnecessary because it is already close to its melting point. Sintering produces thick, necklike bonds between snow grains. The necks are almost as thick as the grains themselves, resulting in a very strong material that sometimes can only be cut with a saw.

Figure 40: This snow wall has been built to shield the tent from the wind. The blocks of drift snow have been cut with a knife. Sintering has made them quite strong and cohesive, allowing the wall to lean but not fall over.

Sintering produces a magical transition. Immediately after the wind stops blowing, the drifted snow is soft and easy to shovel, soft enough for a boot to sink into it several inches (8 cm). But twenty-four hours later, after the sintering is complete, the same snow will be so hard that it is impossible to make a mark on the snow with a boot heel. It will be too hard to be shoveled, but it can be cut into solid blocks with a snow knife or saw (Figure 40). Once the drifts have hardened like this, the snow can no longer be moved by the wind, even if it starts blowing harder than before.

Figure 41: If the snow had not sintered (bonded), it would not be able to hang from the roof like this.

Drifts

The drifting snow piles up in dunes and barchans (crescent-shaped dunes). Even while the wind is depositing the snow, it may also be eroding them at the same time. Fantastic shapes called *sastrugi* (Russian word) can result when these dual processes act in concert.

Changing Snow

The wind creates fantastic drifts and snow shapes.

sastrugi (quyuġlak)
1 meter (about 3 feet)

barchans (aniuvauraq)
2 meters (about 6 ½ feet)

whalebacks (aġviuraq)
10 meters (about 33 feet)

23

It is rare when driving a snowmobile in the Arctic to have a smooth ride. The snow surface is usually covered by dunes and drift deposits that can be 20 to 30 inches high (50 to 76 cm) and which make for a bumpy ride (Figure 42). Depending on how old the drifts are, and how much sintering has taken place, the drifts can be soft or hard. Hard drifts are more common, and much less pleasant to drive over, than soft drifts. A few hours of driving over hard drifts can make it feel like your teeth are going to fall out.

Many types of drift forms have been identified and named. Three of the most common forms in the Arctic are:

- barchans (crescent-shaped dunes with the horns pointing downwind)
- *sastrugi* (Figure 43)
- dunes (Figure 44)

Figure 42: Jon Holmgren maneuvers his snowmobile and sleds over a rough patch of *sastrugi* in Nunavut, Canada.

Figure 43: Jon Holmgren next to a patch of *sastrugi*. These *sastrugi* were originally part of a dune (possibly a barchan) more than 18 inches (46 cm) thick. The wind (blowing from the lower right toward the upper left) has eroded the leading edge of the barchan drift, forming deep flutes and grooves.

Figure 44: Ken Tape is standing on a small whaleback dune that is being deposited by saltating snow grains. The grains are being blown by a stiff wind that is coming from the right side of the picture.

Cornices

The snow saltates along the ground until it drops into a gully, or until it cascades over the edge of bluff. There it comes to rest, piling up through the winter. Eventually, it forms a curling wave of snow, frozen (sintered) in place, that is called a cornice.

The drifting snow fills in gullies and cutbanks. More snow is added during each new blizzard.

WIND
hard snow bank (*sițliq*)
cornice (*mavsa*)
JAN.
DEC.
NOV.
snow-filled gully (*aniuvak*)
24

Generally the snow on the tundra is shallow, often less than 24 inches (60 cm) deep. But in gullies and in the lee of cutbanks, blowing snow is trapped and forms deep drifts, some in excess of 120 inches (305 cm) deep. The drifts build up through the winter, each new blizzard adding snow and extending the drift farther across the gully (Figure 45). In a three-year study of a large drift in arctic Alaska, we found that as few as five wind events were all that was needed to build the drift and deposit 120 inches of snow.

A snow cornice is created by a wind rotor, a small vortex in the wind that forms on the lee side of a ridge or downwind from the edge of a cutbank or gully edge (Figure 46). Saltating snow grains get blown along the snow surface until they are swept over the edge of the bluff or gully. The rotor wind curls the grains down and under. Sintering "glues" the

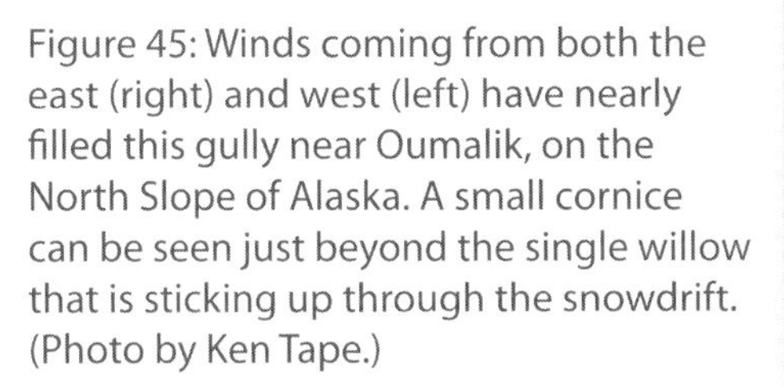

Figure 45: Winds coming from both the east (right) and west (left) have nearly filled this gully near Oumalik, on the North Slope of Alaska. A small cornice can be seen just beyond the single willow that is sticking up through the snowdrift. (Photo by Ken Tape.)

grains together, and soon the snow has mimicked the shape of the rotor. With time, the weight of the cornice will cause it to droop and settle, accentuating the shape even more. In fact, it is not unusual for the cornice to crack a few feet back from the lip. These cracks can be dangerous. They can be deeper than a person is tall, and they can be hidden by a thin layer of snow. An unwary traveler can fall into the crack and get stuck.

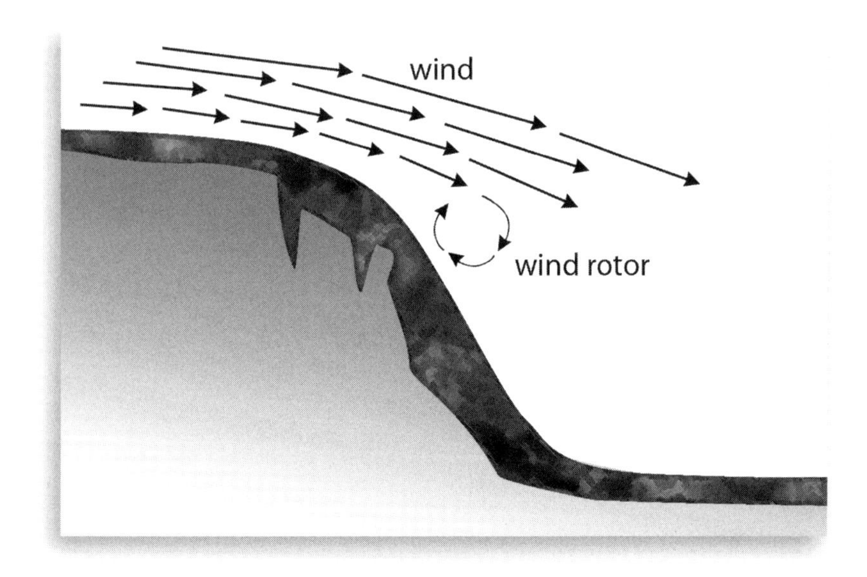

Figure 46: When the wind blows over the edge of a gully, it forms a wind rotor (a horizontal vortex) because it cannot bend steeply enough to follow the surface of the snow. Saltating snow grains follow the wind and get swept into the rotor, which creates essentially a breaking wave of snow, better known as a cornice. Once formed, the snow cornice sags or bends downward under its own weight, accentuating the cornice shape.

Temperature Gradients

The second force that changes the *apun* is a strong temperature gradient. We are all familiar with temperature gradients, often without recognizing it. For example, on a cold day, say –2°F, under our parka and shirt our skin temperature is still 98°F. The temperature drop across our parka might be more than 100°F over a thickness of just a few inches.

Chang
Snow

What is a temperature gradient? Imagine it is cozy and warm inside (fire), but frigid outside (icicles). One side of the wall is warm, the other cold. The temperature gradient is the difference between the warm and cold air. It moves moisture from inside to outside of the house.

25

Temperature gradients are the second of the three metamorphic forces acting on the snow cover. They begin to act in November when it gets cold but the snowpack is still thin. These gradients drive sublimation and condensation of water vapor, producing rapid changes in the size and shape of snow grains. The resulting grains, while different than snowflakes, can be equally beautiful, with elegant features and sharp geometric angles.

What is a temperature gradient? Think of a house in winter. The furnace keeps the inside at 68°F (20°C) even when it is –40°F (–40°C) outside. That is a 108°F temperature drop, which is pretty big. If the wall is 8 inches thick (20 cm), then there is a temperature gradient of 108°F per 8 inches, or about 14°F per inch of wall (3°C per cm).

The snow cover is almost as good an insulator as a wall insulated with fiberglass. It is not uncommon to find the temperature at the base of the snow 30° to 40°F warmer than the top of the snow because of this insulating property, and for these temperature differences to last for weeks at a time. These conditions produce strong temperature gradients and rapid metamorphism.

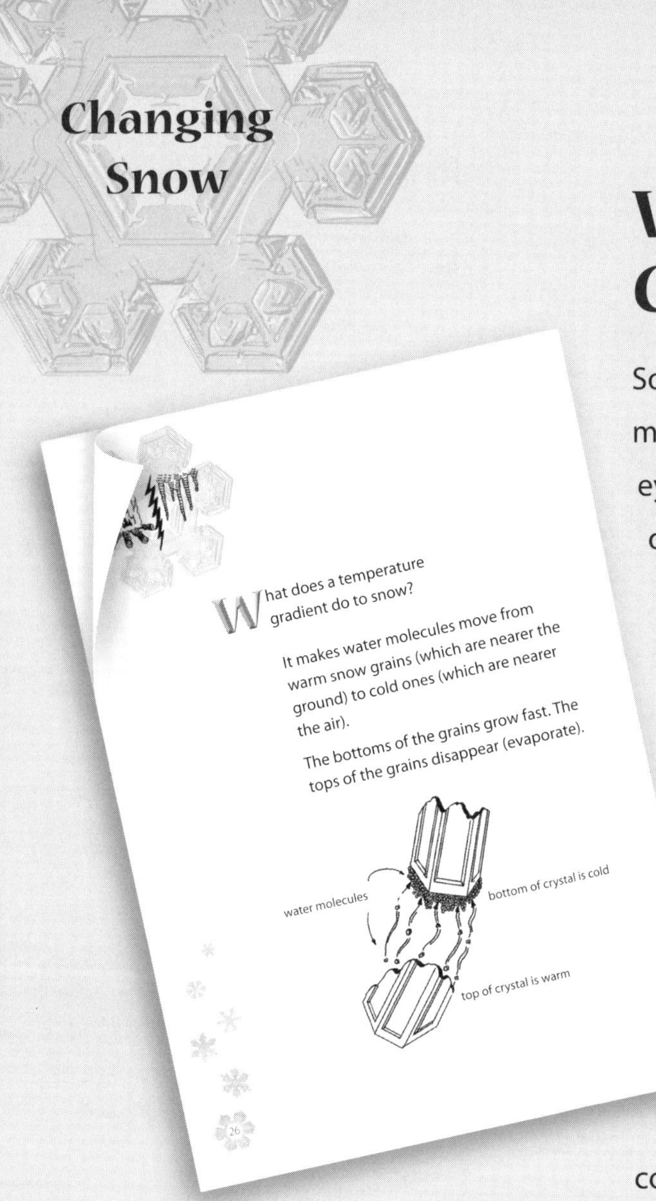
What does a temperature gradient do to snow?

It makes water molecules move from warm snow grains (which are nearer the ground) to cold ones (which are nearer the air).

The bottoms of the grains grow fast. The tops of the grains disappear (evaporate).

Vapor Sublimation and Condensation

So how does a temperature gradient alter the snow? Unlike melting and wind drifting, which we can see acting with our eyes, temperature gradient metamorphism is more mysterious because we cannot see the process in action, only the end result. It works like evaporation. Set a pan of water out in hot weather and all of it may disappear in a day, the water going off one molecule at a time, the actual process not visible to us. The warmer the water, the faster the evaporation rate. In the same way vapor sublimation (evaporation from ice) and condensation works on the *apun* to grow beautiful and complex grains.

A Japanese scientist named Yosida coined the term "hand-to-hand growth" to describe how through this process water molecules move upward in the snowpack, evaporating from the tops of grains and condensing on the bottoms of the next higher grains. The process is a little like a bucket brigade!

To understand how sublimation leads to snow metamorphism, and how it is connected to temperature gradients, it is useful to imagine that you are tiny, smaller than a grain of snow, and that you are crawling through the snow cover in the tiny open cavities between the grains. These cavities are called pores. Your world might look something like the black-and-white drawing in Figure 47. Because the ground beneath the snow cover is warmer than the air above it, heat will flow upward from the bottom to the top of the snowpack. The snow grains, being made of ice, conduct heat nicely, but the pores, being air-filled, do not. Standing in a pore space, with your feet on one grain and your hands touching the grain above your head, you would notice your feet were warm and your hands were cold. Water molecules sublimating from the lower grain would travel up across the pore space and condense (be deposited) when they came in contact with the colder grain over your head. The bottom of the upper grain would grow, while the top of the lower grain would shrink. In fact, that is exactly what happens, as Figures 47 and 48 show. Snow scientists call this type of snow metamorphism temperature gradient or dry snow metamorphism.

A Japanese snow scientist named Yosida studied how snow grains grew and metamorphosed under a temperature gradient. Yosida was one of the first to recognize that the source of moisture for one growing grain was often a loss of moisture by the grain immediately below it. He called this a "hand-to-hand" process, likening it in some ways to a bucket brigade moving pails of water to a fire. The loss (sublimation) leads to rounded edges and large featureless facets (like smooth panes of glass). The gain (condensation) leads to sharp edges and ornate and intricate growth features like striations (Figure 49).

The larger the difference in temperature between the upper and lower grains, the more rapidly the crystal will grow, and to some extent, the more exotic the forms that will develop. Striated, cupped, skeletal, and scrolled crystals develop when the gradient is very strong.

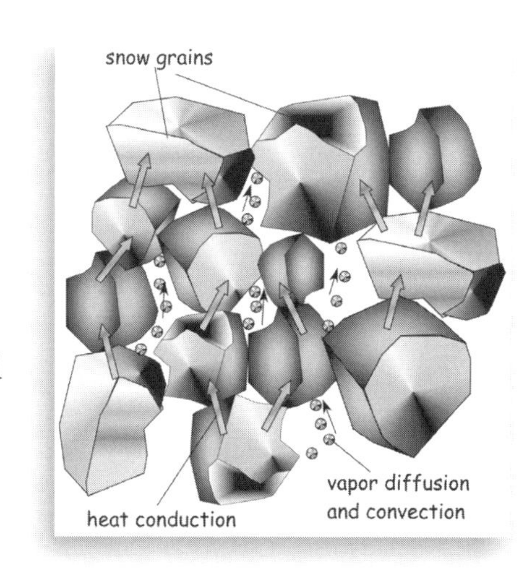

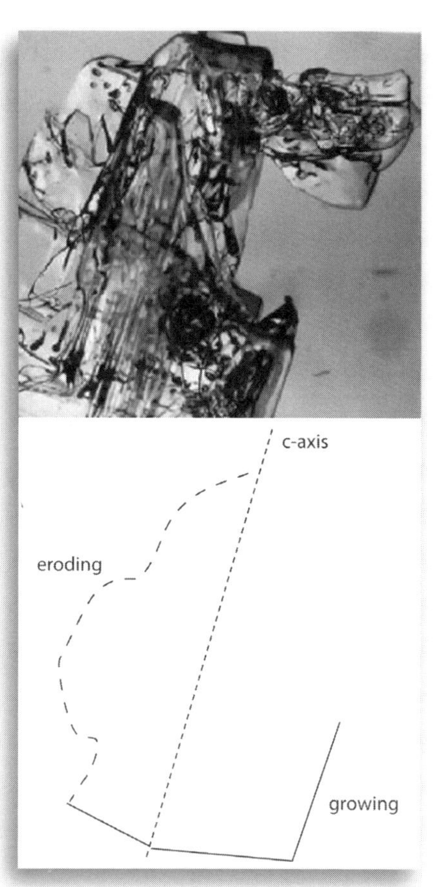

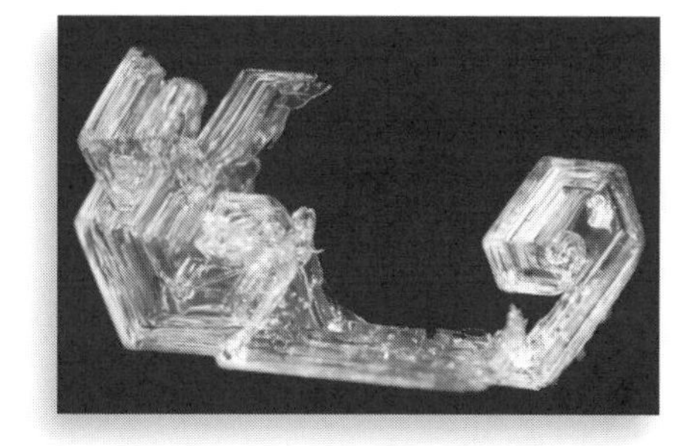

From top:

Figure 47: A schematic showing snow grains and pore spaces. The arrows indicate pathways through ice grains. These are pathways along which heat readily flows. The small spheres indicate water vapor sublimating from ice grains and migrating upward through pore spaces. The water vapor will sublimate from the tops of grains and condense on the bottoms of higher grains.

Figure 48: A snow grain from the base of the snowpack. The bottom and right edges are sharp and show evidence of recent growth. The top edges are eroded and show signs of recent sublimation.

Figure 49: A large (about ¾ inch or 2 cm long) grain from the base of an arctic snowpack showing skeletal growth, cup-shaped segments, and beautiful scrolling.

Depth Hoar Crystals

The hand-to-hand process of sublimation and condensation is simple, but it can give rise to a wide range of beautiful grain forms that are called depth hoar (or *pukak* in Iñupiaq) because they are usually found near the bottom of the snow cover.

Changing Snow

solid-type depth hoar

0.3 mm

depth hoar cups

3 mm

scrolled depth hoar

3 mm

These fast-growing crystals develop beautiful geometric shapes. They are called depth hoar or *pukak*.

27

For three years I ran an experiment on temperature gradient metamorphism in the subarctic snow cover. During that time I spent countless hours lying on my side looking at the snow grains with a magnifying glass. I believe that by the end of the experiment, I could almost see water molecules moving through the snowpack by the hand-to-hand process. Perhaps I did . . . perhaps not. What I did do is get to see some very beautiful depth hoar crystals. In Figure 50 I am once again getting up-close and personal with *apun*, in this case to photograph the depth hoar.

Most people have heard of hoar frost. Fewer people have heard of depth hoar. The word *hoar* indicates a large feathery type of crystal (not the disreputable ladies of the night, which is spelled differently). Since these feathery crystals are more prevalent at the base of the snowpack, they are called depth hoar.

There is a very regular progression of crystal forms as large depth hoar crystals grow and develop. First, small, squat solid crystals grow (Figure 51a). These look like the cut-glass crystal of a chandelier. Next, the squat crystals start to grow downward and outward into bell-like cups (Figure 51b). The cup edges develop stepped and striated edges. Finally, exotic scrolled crystals (Figure 51c) replace the cup crystals.

Figure 50: Matthew Sturm examines depth hoar in arctic Canada. (Photo by Henry Huntington.)

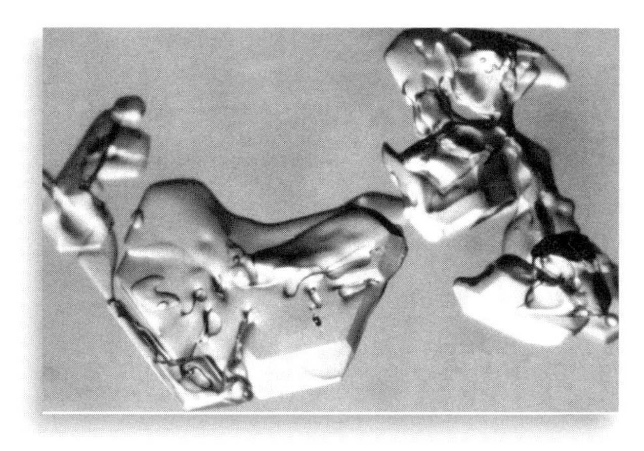

Figure 51a: This grain shows hexagonal symmetry and sharp facets on the bottom edges, but the top has been heavily eroded and rounded by sublimation.

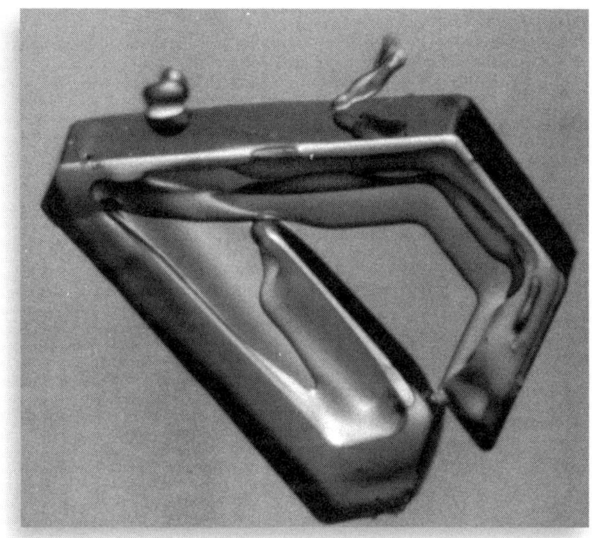

Figure 51b: This unusual grain looks triangular, but actually has hexagonal symmetry. Again, condensation has left sharp geometric edges.

Figure 51c: This is a classic depth hoar grain resulting from temperature gradient metamorphism. In addition to the beautiful hexagonal symmetry and sharp edges on the bottom, it exhibits well-developed striations that look like air channels in the face of the grain.

Pukak

In the *Apun* drawing, the hunter is standing on a layer of strong wind-slab snow. His boots barely sink in. At the bottom of the snowpack, there is a layer of very large ornate depth hoar grains that have grown due to strong temperature gradient metamorphism. This layer is brittle and weak. Voles and lemmings like to live in this layer because it is easy to tunnel through. In Iñupiaq this layer is called *pukak*.

The first snow of the year (*apivaalluqqaaġniq*) filled in the holes between tundra tussocks. It has been there since October. As more snow has fallen on top of it, the bottom layers have gotten warmer. Driven by a temperature gradient, the snow grains have now grown into depth hoar crystals. This layer is also called *pukak* in Iñupiaq.

Pukak

28

Temperature gradient metamorphism produces the most extreme changes at the base of the snow cover. The basal snow has been in place the longest, and it is also warmest because it is nearest the ground. Both factors are conducive to Yosida's hand-to-hand vapor transport. By the end of the winter, the grains at the base of the snow can be ½ inch (13 mm) to ¾ inch (20 mm) long and extremely ornate (Figure 52). This coarse-grained basal layer is called *pukak* in Iñupiaq.

Figure 52: This depth hoar crystal is almost 1 inch long (25 mm) from the base of the snow.

Playing on Depth Hoar

What a wonderful crystal playground it would be if the *pukak* layer was so large we could climb around on it like a jungle gym.

Figure 53: These depth hoar crystals were found at the bottom of the snow cover in Nunavut, Canada. Imagine being small and playing in this crystal playground! (Photo by Jon Holmgren.)

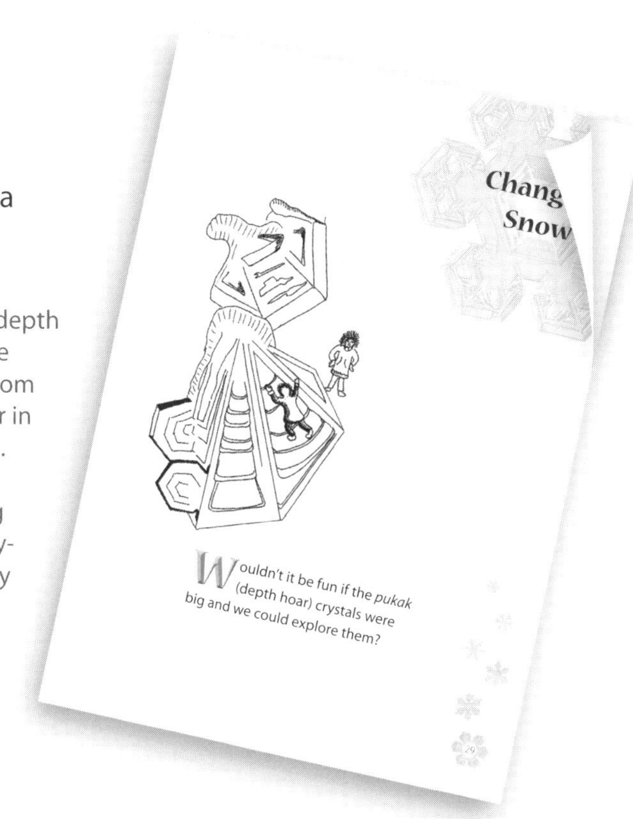

Observing Depth Hoar

While climbing on *pukak* is not possible, we can explore the crystals with a hand lens. It is easy. Put some crystals in your hand and look at them outdoors where there is good lighting. Better still, spread them on the sleeve of your jacket (better if it is dark-colored) and look at them there.

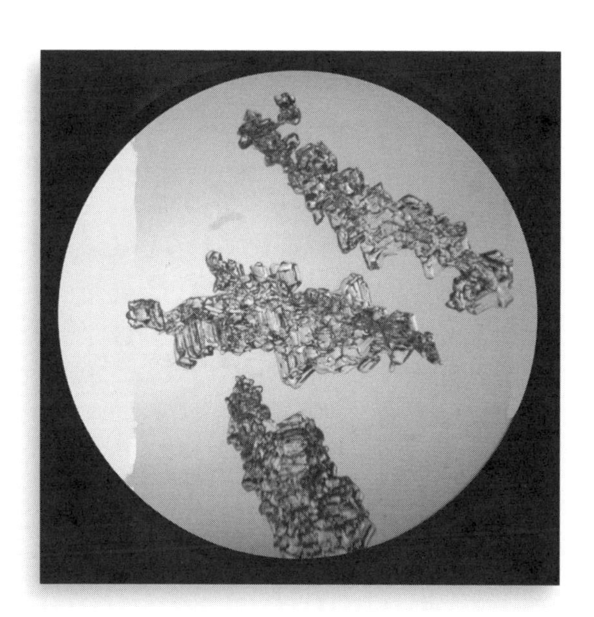

Figure 54: Several chains of depth hoar can be seen in the magnifying glass here.

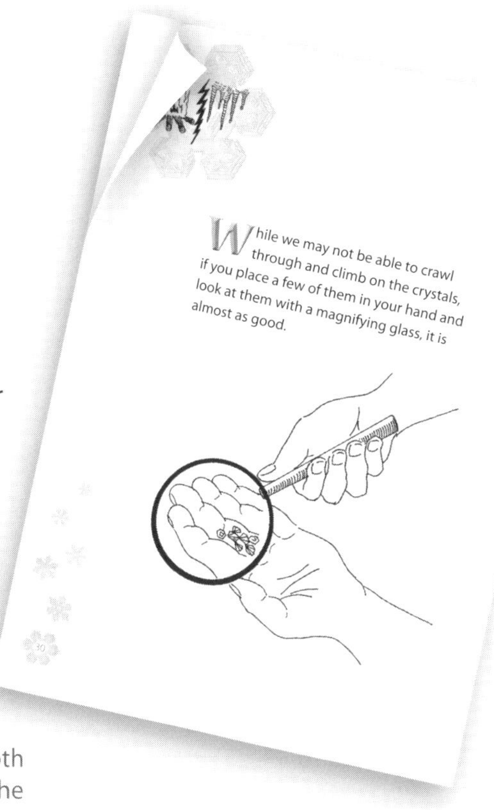

Heat

The third force that changes the *apun* is heat. This heat can come from the sun, from a warm weather front, or from rain falling on the snow. This type of metamorphism happens in spring. With enough heat, the whole snowpack will melt away, ending the snow year, but often, just a little heat is introduced, changing the crystals to yet another form.

Chang
Snow

Heat

Finally a day comes when the sun comes out bright and strong. The air is warm and still. There is a strange sound, one that hasn't been heard since last spring: drip, drip, drip.

31

The photograph on the left (Figure 55) was taken in the Sierra Nevada Mountains in July. The air temperature was almost 90°F (32°C)! Not surprisingly, the snow was melting and an icy creek was pouring out from the snow. The photograph on the right (Figure 56) was taken in the Alaska Arctic. The sun is shining through diamond dust, so we know it is cold. The sun has little power to melt snow just at this time, but as spring comes on, its power will increase, and it too will soon supply enough heat to melt the snow. Unlike depth hoar, the result of adding this heat to the snow is to produce beautifully rounded and smooth grains.

Figure 55 (left): Melting snow.

Figure 56 (below): The weak arctic sun in winter.

Melt Clusters

The basic grain shape produced by adding heat to *apun* is round. The amount of heat added controls just how rounded the grains become, and how thick the bonds connecting the grains will be. These bonds should not be confused with the bonds produced by sintering, which happens without any liquid water being present.

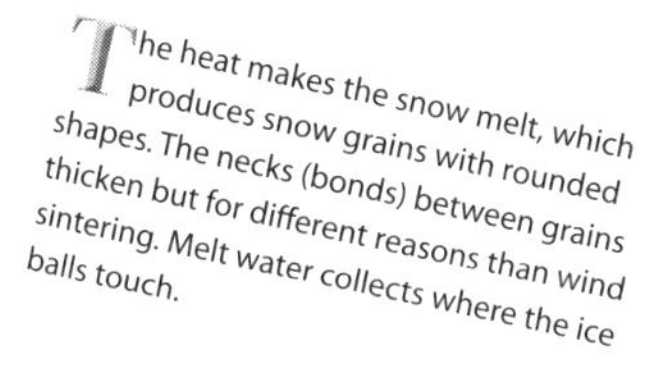

The heat makes the snow melt, which produces snow grains with rounded shapes. The necks (bonds) between grains thicken but for different reasons than wind sintering. Melt water collects where the ice balls touch.

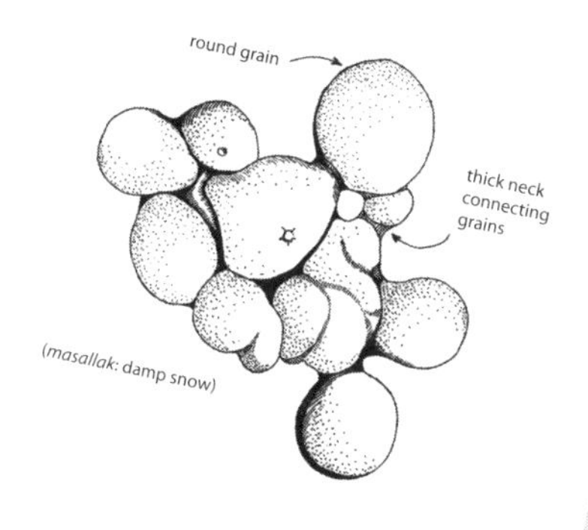

Melting is another type of snow metamorphism, and it is easy to observe. Bring a cup of snow inside and watch as it melts. The new snowflakes first lose their points, then their arms, then they round into small wet spheres like those shown in Figure 57, then the area where the spheres touch gets wet and widens until the whole mass, if it refreezes, is incredibly strong.

Figure 57: Melt grain clusters. Note the rounded grains, the thick necks connecting grains, and the refrozen water filling the pore spaces between grains.

Percolation Columns

Everyone knows what an icicle is. Well, when the snow is heated and melts at the surface, it will form icicles inside the snow! These are called percolation columns and they can be as big as a person's arm. The reason meltwater forms these features is that the water does not like to spread out into the cold snow through which it is flowing. The exception to this rule is at the boundaries between snow layers, which is why the columns have ribs.

Chang
Snow

percolation column

ice layers

ice layer

ice

If there is a lot of melting, then melt water will "percolate" down into the snow and spread out along snow layers, where it will refreeze and make ice layers. A surprising trick is that the melt water can make several ice layers at one time, the layers connected by knobby pipes called percolation columns. Can you find these?

33

The sun warms the snow surface and produces meltwater there. The water finds a weak spot in the surface snow layer and drains downward. Because the snow into which the water is flowing is cold, some of the water freezes, but the water keeps draining down the outside of this internal icicle. Sometimes, the water "steps over" to find a new weak spot in the layers, producing several ice layers and more than one percolation column. The percolation process is easy to observe. On a hot day when the snow is melting, spray the snow with water colored with food dye. Cut into the snow with a shovel and you can see where the water is percolating.

Figure 58: Two percolation columns that were excavated from a subarctic snowpack near Nome, Alaska. In several places, the columns have spread out along snow layers then refrozen, creating the ridging or ice ribs.

Firnspiegel

When heat melts the surface of the snow, it can produce a bright, shiny layer that reflects sunlight and is very slippery.

Sometimes just a little melting will take place right at the surface. Perhaps it will be warm enough for mist or even rain. The snow surface will become hard and shiny. In Iñupiaq this is called:

siḷḷiġruaq

The Germans have a nice word for this:

firnspiegel

It means ice mirror.

34

In April 2007 we were traveling east across the Barrenlands of Canada. As we moved east, we came to an area where a week before our arrival there had been freezing rain. This had produced an ice crust on the top of the snowpack. The rainstorm had turned into a snowstorm toward the end, so the icy surface, the *firnspiegel*, was covered by a little snow. But after a few miles, we came to an area where the wind had swept the new snow from the higher bumps (Figure 59). A little farther on, even more of the icy snow was exposed. Walking became treacherous because the icy layer was strong enough to support our weight but was so slick it was almost impossible to get any traction on. We all fell several times before we learned to step only in places where the *firnspiegel* was covered with the new snow. Iñupiat hunters have told me that they have seen caribou fall when slipping on surface ice layers (in Iñupiaq, *siḷḷiġruaq*) in the same way. As the climate warms, there is some indication that adverse conditions like this may increase in frequency.

Figure 59: *Siḷḷiġruaq*, also known as *firnspiegel*, peeking through a layer of new snow in Northwest Territories, Canada. It is so shiny, it reflects the sun like a mirror.

Slush

Everyone knows what slush is, but how many people have ever actually looked at the floating crystals?

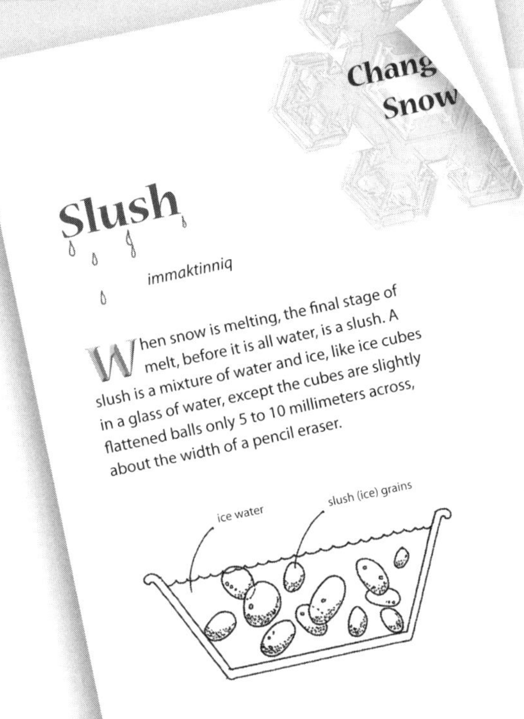

Chang
Snow

Slush

immaktinniq

When snow is melting, the final stage of melt, before it is all water, is a slush. A slush is a mixture of water and ice, like ice cubes in a glass of water, except the cubes are slightly flattened balls only 5 to 10 millimeters across, about the width of a pencil eraser.

ice water

slush (ice) grains

35

If above-freezing temperatures continue long enough (maybe a few days), a slush will form. A slush consists of oblate ice spheroids (slightly flattened balls of ice) (Figure 60) floating in ice water. At any time, the weather can turn cold and "refreeze" the melting grains, halting the changes before they have produced ice spheres or slush. When that happens, the result is what are called melt grain clusters. The melt clusters look like a bunch of grapes or glass globes fused together.

Figure 60: Slush consists of ice grains (oblate spheroids) suspended in ice water.

Snow and People

Snow Knives

Tools are the hallmark of humans . . . and the snow knife is one tool that shows how important *apun* has been to arctic people.

Snow and People

The arctic snow cover has been important to northern people for centuries. Before the Iñupiat had metal, they made snow knives (panak or saviuraqtuun) from bone and ivory. Some bone knives are more than 600 years old! Today, snow knives continue to be used, but they are made from steel. They are still used to cut and shape snow, because even today, working with *apun* means survival in the North.

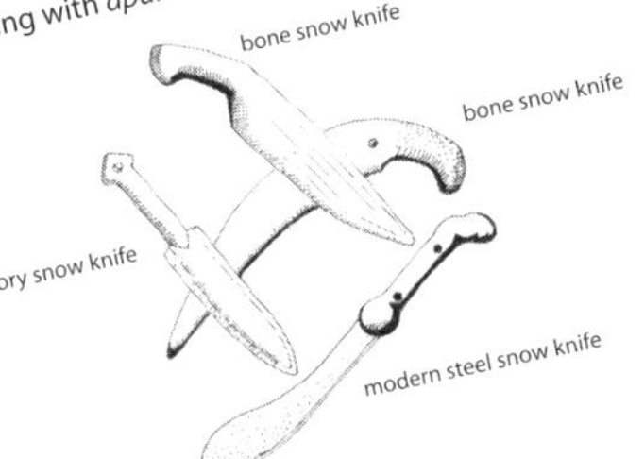

37

I became interested in snow knives during a long winter trip in Canada. Snow knives have been in use for thousands of years. Knives have been dug up in archeological sites that are made of bone, antler, or ivory. Metal knives came into use in the 1700s when the whalers came to the Arctic, and knives are still in use today.

Counterclockwise:

Figure 61: Two modern snow knives. The one on the left is from Baker Lake, Nunavut. I bought it from an Inuit elder and repaired the handle. The one on the right is made from a machete blade following a traditional pattern.

Figure 62: An Iqaluit Inuit bone snow knife.

Figure 63: This still picture was taken by Robert J. Flaherty, who created the silent film *Nanook of the North* (1922). It shows an Inuit man trimming snow blocks with a bone snow knife.

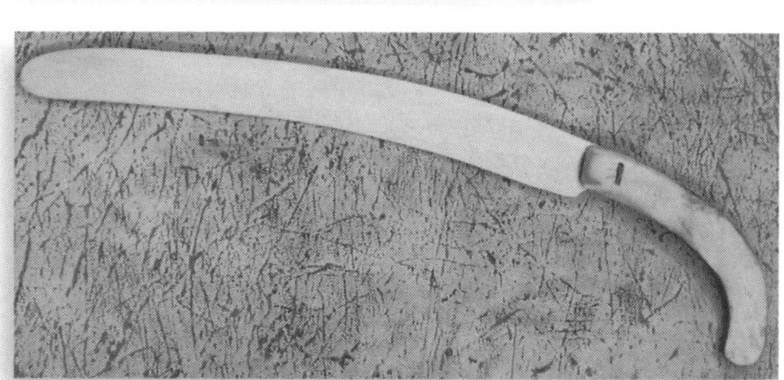

Snow Houses

Nothing illustrates the importance of *apun* to humans more than the igloo—the home built of snow.

As one might imagine, living in an igloo takes practice. Typically, a whole family lived in an igloo, perhaps four to six people. Things had to be well organized or chaos would reign. As the figures below (based on *Our Arctic Way of Life: The Copper Inuit* by Doreen Bethune-Johnson) indicate, the inside of an igloo was organized in a way that allowed for all the various life activities associated with hunting and subsisting on the land, as well as living in a snowy environment. Note the critical component of the lamp for cooking and drying clothing over by the table and the drying rack.

One use of snow knives is to make shelter. Good snow houses (*aniquyyaq*) must be built from strong, deep snow (*aniuvak*). The snow blocks must be cut with skill so that the shelter won't blow down.

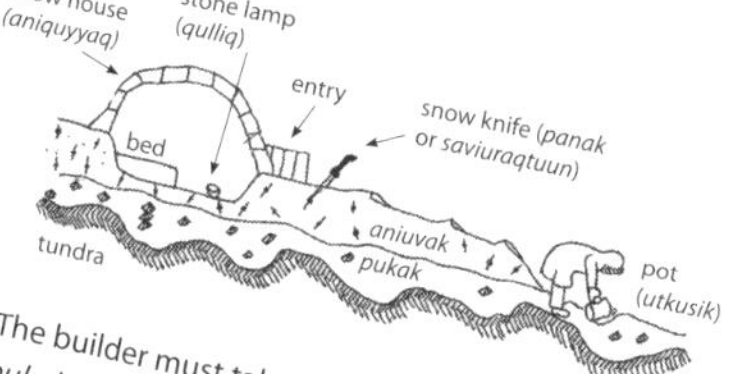

The builder must take care not to dig into the *pukak* at the bottom of the snow or else the wind will blow through the snow into the house. The loose *pukak*, however, is good for filling a pot to melt snow for water because it is weak and scoops up easily. The house blocks are glued together with loose snow that sinters (remember?) one block to another.

38

Figure 64: Frobisher's 1576–1578 expedition to Baffin Island in the Canadian Arctic made this engraving of an Inuit igloo village, one of the first times snow houses were brought to the attention of the Western world. Notice the windows all facing south, where they will get the most light.

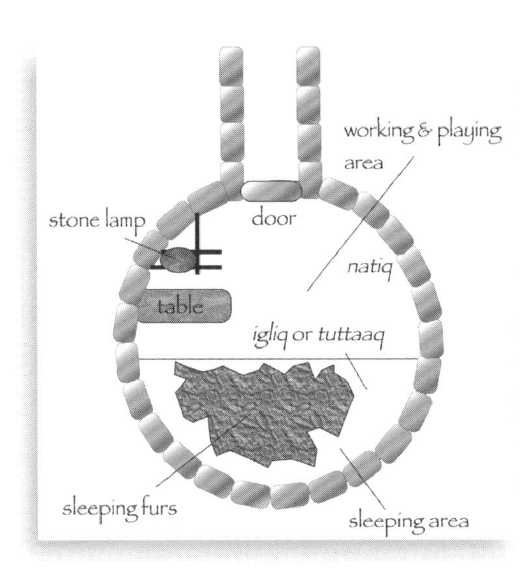

Figure 65: A sketch of the inside of an igloo.

Snow and Disappearing Sea Ice

Apun is critical to humans in another way: it plays a major role in arctic climate warming. This is particularly true on the sea ice, where the snow can control whether the ice is thick (and will last through the summer) or thin (and could melt away).

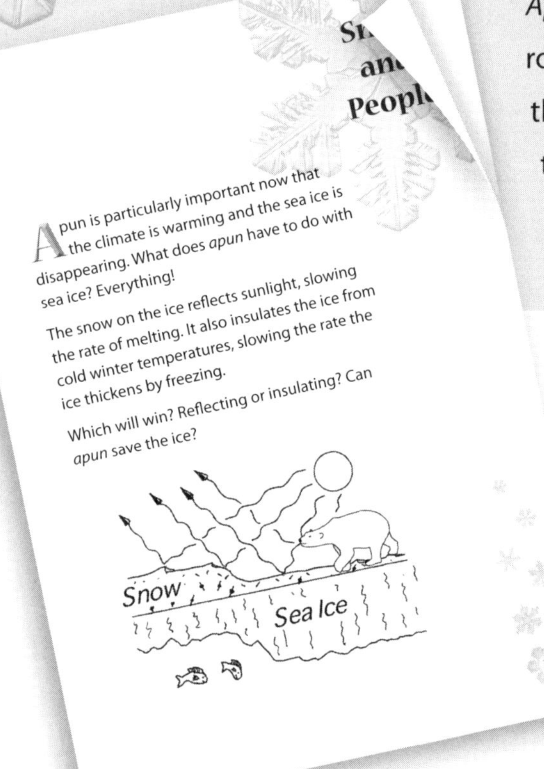
Sn an Peopl

Apun is particularly important now that the climate is warming and the sea ice is disappearing. What does *apun* have to do with sea ice? Everything!

The snow on the ice reflects sunlight, slowing the rate of melting. It also insulates the ice from cold winter temperatures, slowing the rate the ice thickens by freezing.

Which will win? Reflecting or insulating? Can *apun* save the ice?

39

The snow insulates the sea ice and reflects sunlight. How thick the snow cover is, and when it arrives and melts, plays a major role in determining whether the ice grows thick during the winter and/or whether it melts rapidly during the warm weather in spring. The snow may just be the "wild card" in whether the summer sea ice survives over the next few decades.

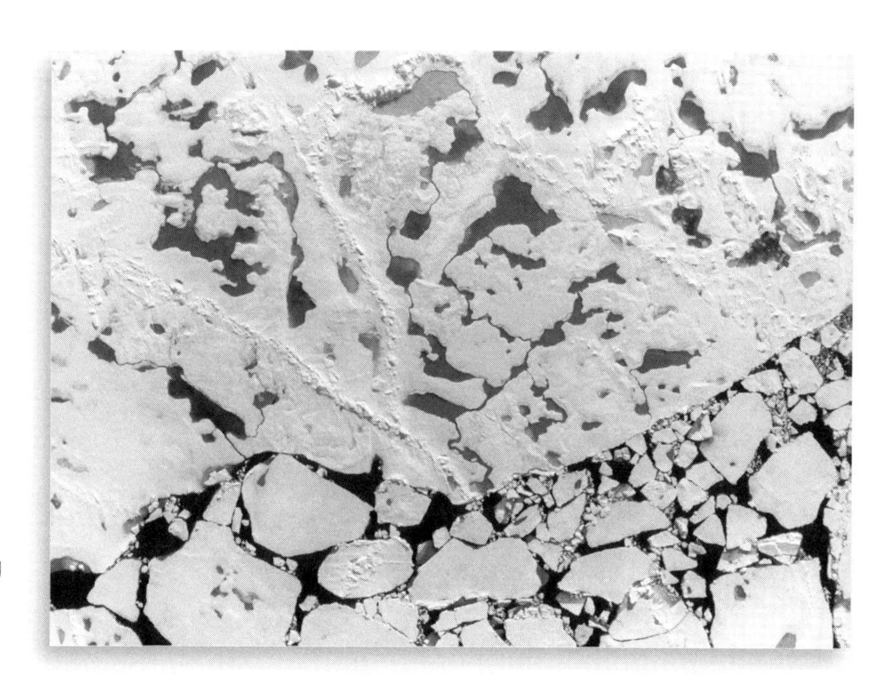

Figure 66: This aerial view of the sea ice comes from about 300 miles north of Barrow. It shows snow-covered ice (white), freshwater melt ponds on the ice (gray), and ocean water (black). The white ice reflects almost 90% of the incoming solar radiation, but the ocean absorbs most of the solar radiation. The loss of highly reflective sea ice (replaced by ocean water) due to climate warming is helping to amplify and reinforce the effects of warming in the Arctic.

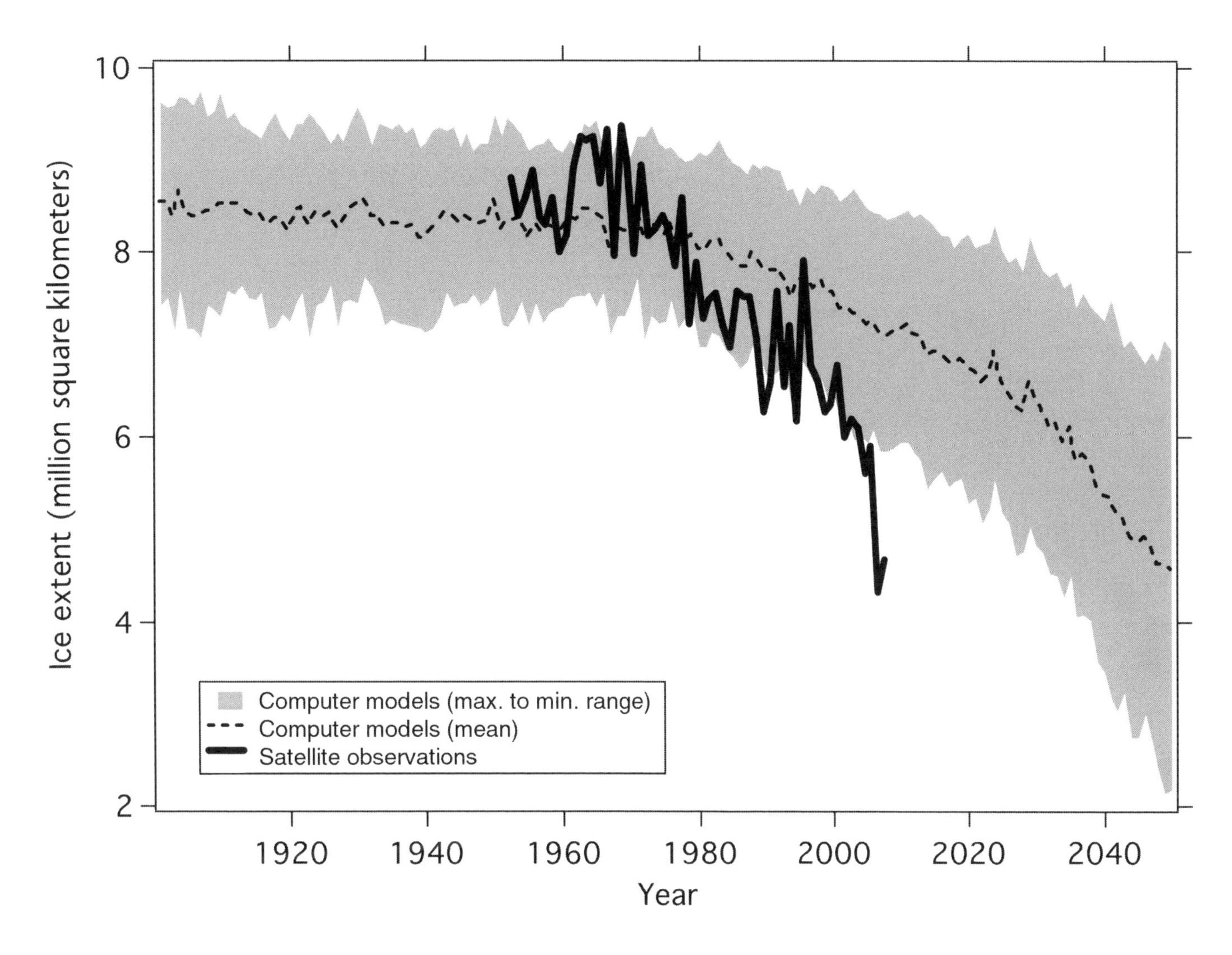

Figure 67: This graph tells the story: the summer sea ice extent (how much sea ice remains at the end of the summer) has been declining rapidly since the 1960s. The strongest decline ever was in 2007, and in 2008 and 2009 nearly as much melted. The most worrisome aspect of this decline is that none of the global climate models (all the other lines) have predicted as rapid a decline as we have actually observed.

The story comes full circle back to life. Living in the Arctic means living with *apun*, the arctic snow cover, for plants, animals, and people. The scene above illustrates just how important the snow, and knowledge of the snow, was and still is for the people of the Arctic.

Apun is life!

Iñupiaq Glossary

Agniqsuq. a full-on dry snow blizzard, during which it is hard to see anything because it is a whiteout (*quvyugaġnaq*). (See also *piqsiqsuq*: a wet snow blizzard.)

Aġviuraq. a whaleback drift (possibly spelled *avoorik*).

Aniquyyaq. a snow house or igloo.

Aniu. packed snow.

Aniuvak. hard-packed snow in a gully. A big drift in the lee of a cabin.

Aniuvauraq. a drift with a sharp downwind side and a smoother upwind side; a crescent-shaped drift called a barchan.

Anuġi. wind.

Apivaalluqqaaġniq. first snow of the year in October.

Apun. the snow cover, a general term for snow on the ground.

Augniqsraq. a patch of tundra from which snow has melted.

Auktuq. melting snow.

Immaktinniq. slush; very wet snow, as in an area where river water flowed into the snow (a snow swamp).

Irriqutit. diamond dust or ice crystals in the air (which means a cold spell is coming).

Masallak. snow damp enough to stick, as in making snowballs; also spelled *masayyak*).

Mavsa. an overhanging snow cornice (in Kotzebue: *mapsa*).

Miñŋuq. a little jumping beetle, sometimes red (comes from the word *minŋiq*, which means to jump up or over).

Natiġvik. low blowing snow, with drift no higher than the knee.

Natiqłit. even lower drifting snow than *natiġvik*.

Nutaġaq. fresh powder snow.

Nuturuk. good material to make a snow house, firm yet not too hard.

Panak. a snow knife (Canadian: see *saviuraqtuun*).

Piqsiqsuq. a wet snow blizzard.

Pukak. coarse, large ornate depth hoar crystals making up a layer at the base of the snow cover.

Qannik. a snowflake.

Qanuġinniŋa siḷam. the drop in temperature across a house wall or the snowpack; a temperature gradient.

Qayuqłak. an anvil-head drift or *sastrugi*.

Qimuagruk. a long, built-up snowdrift similar to a whaleback (*aġviuraq*), useful for navigation.

Qulliq. a stone lamp

Saviuraqtuun. a tool (i.e., knife) used by a man, possibly a snow knife (see *panak*).

Siḷḷiġruaq. slippery icy snow conditions, as a surface ice layer.

Siḷḷiq. hard icy snow above *pukak*, sometimes the product of a rain-on-snow event.

Siḷḷiqsruq. old icy snow, extra-hard. So hard no snowmachine or animal tracks are left on it. Hard to dig with a shovel. The snow surface can be a lot like lake ice. Looks like *firn-spiegel* (ice mirror).

Utkusik. a cooking pot.

Uunnaq. heat.

Index

Q

R

S